Mohamed Abdel-Raheem

Biopesticides

AF755959

Mohamed Abdel-Raheem

Biopesticides

Noor Publishing

Imprint
Any brand names and product names mentioned in this book are subject to trademark, brand or patent protection and are trademarks or registered trademarks of their respective holders. The use of brand names, product names, common names, trade names, product descriptions etc. even without a particular marking in this work is in no way to be construed to mean that such names may be regarded as unrestricted in respect of trademark and brand protection legislation and could thus be used by anyone.

Cover image: www.ingimage.com

Publisher:
Noor Publishing
is a trademark of
Dodo Books Indian Ocean Ltd. and OmniScriptum S.R.L publishing group

120 High Road, East Finchley, London, N2 9ED, United Kingdom
Str. Armeneasca 28/1, office 1, Chisinau MD-2012, Republic of Moldova, Europe
Managing Directors: Ieva Konstantinova, Victoria Ursu
info@omniscriptum.com

Printed at: see last page
ISBN: 978-620-8-87082-9

Copyright © Mohamed Abdel-Raheem
Copyright © 2025 Dodo Books Indian Ocean Ltd. and OmniScriptum S.R.L publishing group

Biopesticides

By

Prof. Dr. Mohamed Abdel-Raheem

(2025)

Biopesticides

Prof. Dr. Mohamed Abdel-Raheem Ali Abdel-Raheem

Professor of Entomology (Biological Control)

Pests & Plant Protection Department

Agricultural and Biological Research Institute

National Research Centre, Cairo, Egypt.

Contents list

INTRODUCTION

Biopesticides are certain types of pesticides derived from such natural materials as animals, plants, bacteria, and certain minerals.

Biopesticide is comparatively safer strategy than the use of chemicals or chemical derived pesticides posing low risks to humans and the environment.

The major advantage of the use of biopesticide for crop pest's management is environment safety along with their host specificity. Besides, the associated costs for the development and the registration of the biopesticide are comparatively lesser than that of the chemical pesticides. Besides, due to the continuous evolution of the resistance to chemicals and legislations being more restrictive, there is a need for intensive research on the biopesticide development both from the private and the public sectors.

Nevertheless, there are several issues with the development and use of biopesticides in agriculture. Some of the major issues include development of the resistances in the target organism, evolution of the pests, shifts in the pest type, changes in the doses, evaluation of new species, etc.

Further research using genetic engineering tools along with synthetic biology will be important to determine the basis of the gene actions and the evolution of the resistance of the target organism. CRISPR/Cas9 is an important technique that has the potential for the development of genetically modified biological control organisms.

The use of genome sequencing data along with proteomics data will help to identify genes, gene regulatory actions and has potential to explore the genetic basis of resistance. Further research on the chemical modes of actions of biopesticide can resolve issues of host complexity and functionality of the biopesticides in different targeted and non-targeted organisms.

Research on the precision application of the biopesticide by utilizing big data acquired through different sensors and the drone-based system might warrant the effective utilization of biopesticides. Reliance on a few strategies for pest management cannot warrant the successful crop production while a holistic approach of compatible measures is important for successful pest management.

A successful biopesticide development can only be obtained through interdisciplinary research involving genetics, genomics, molecular biology, biochemistry, agronomy, plant pathology, physiology, ecology, and social sciences.

The term *biopesticides* defines compounds that are used to manage agricultural pests by means of specific biological effects rather than as broader chemical pesticides. It refers to products containing biocontrol agents – i.e., natural organisms or substances derived from natural materials (such as animals, plants, bacteria, or certain minerals), including their genes or metabolites, for controlling pests. According to the FAO definition, biopesticides include those biocontrol agents that are passive agents, in contrast to biocontrol agents that actively seek out the pest, such as parasitoids, predators, and many species of entomopathogenic nematodes.

Thus biopesticides cover a wide spectrum of potential products that can be classified as follows:

Microbial pesticides and other entomopathogens: pesticides that contain microorganisms, like bacteria, fungi, or virus, which attack specific pest species, or entomopathogenic nematodes as active ingredients. Although most of these agents attack insect species (called entomopathogens; products referred to as bioinsecticides), there are also microorganisms (i.e., fungi) that control weeds (bioherbicides).

Plant-Incorporated Protectants (PIPs): these include pesticidal substances that are produced in genetically modified plants/organisms (GMO) (i.e., through the genetic material that has been incorporated into the plant).

Biochemical pesticides: pesticides based on naturally occurring substances that control pests by non-toxic mechanisms, in contrast to chemical pesticides that contain synthetic molecules that directly kill the pest. Biochemical pesticides fall into different biologically functional classes, including pheromones and othersemiochemicals, plant extracts, and natural insect growth regulators.

Biopesticides generally have several advantages compared to conventional pesticides (Kaya and Lacey 2007, Kaya and Vega 2012). While chemical pesticides are responsible for extensive pollution of the environment, a serious health hazard due to the presence of their residues in food, development of resistance in targeted insect pestpopulations, a decrease in biodiversity, and outbreaks of secondary pests that are normally controlled by natural enemies, biopesticides, in contrast, are inherently less toxic to humans and the environment, do not leave harmful residues, and are usually more specific to target pests.

Often they affect only the target pest and closely related organisms, substantially reducing the impact on non-target species. A further advantage of most microbial pesticides is that they replicate in their target hosts and persist in the environment due to horizontal and vertical transmission, which may cause long-term suppression of pest populations even without repeating the application.

Since the use of the biopesticides is markedly safer for the environment and users, and more sustainable than the application of chemicals, their use as alternatives to chemical pesticides, especially as components in Integrated Pest Management (IPM) strategies, is of growing interest. Several biopesticides of the different classes have proved to be very effective in controlling potato pests; however, there are certain disadvantages associated with their use that have prevented them from being used on a wider basis in potato production today.

The very high specificity of the products might be a disadvantage when a complex of pest species needs to be controlled. Since biopesticides often contain living material, the products have reduced shelf lives. Also, their efficacy is often variable due to the influence of various biotic and abiotic factors.

For using biopesticides effectively, users need to have good knowledge about managing the particular pests or pest complexes. Due to limited commercial use (niche products) biopesticides often are developed by research institutions rather than by the traditional pesticide industry. While effective active ingredients have been discovered, products might lack appropriate formulation for efficient field use.

A broader set of perspectives in the design and launch of a biopesticide would be helpful. Farmers consider biopesticides often as an alternative to a chemical pesticide, in which the active ingredient is thought to be synthetic, having a similar mode of action to the chemical pesticide. But the truth is that biopesticides differ in their modes of action from conventional chemical pesticide considerably; their modes of action are almost always specific.

Therefore, using biopesticides efficiently requires specific user knowledge on the agent and the target pest for optimizing application time, field rates, and application intervals. Biopesticides should not be considered as a one-for-one replacement of chemical pesticides.

As pesticides in general, biopesticides need to be approved and registered as such in most countries before they can be used, sold, or supplied. Since biopesticides pose fewer risks than conventional pesticides, authorities generally require fewer data for their registration. For example, the Environmental Protection Agency (EPA) in the USA often registers new biopesticides in less than a year, compared with an average of more than 3 years for conventionalpesticides.

However, in some cases it is difficult to determine whether a product meets the criteria for classification as a biopesticide, and the decision by local agencies might vary depending on the regulations in each country. There might be specific requirements pertinent to the different categories of biopesticides.

EPA defines three categories for biopesticides but the two main ones are:

Microbial

- Fungi
- Bacteria
- Nematodes
- Viruses

Biochemical

- Insect Pheromones
- Plant extracts, oils & botanicals
- Plant growth regulators
- Insect growth regulators

The use of biopesticides is rising and consumers are becoming more aware of the hazard that conventional pesticides can cause to the environment as well as to human health. This has lead to governments putting more pressure on farmers to use less traditional chemicals and regulators to approve fewer pesticides. Additionally most hazardous pesticides are being phased out and the availability of broad spectrum chemicals is reducing, which gives an opportunity for alternative technologies.

There are many benefits of using biopesticides as part of a crop protection strategy. Their low re-entry and pre-harvest intervals after application means crops can be protected right up to harvest which is also helped by their lack of residue.

Biopesticides have lower environmental impact as they are usually less toxic than synthetic pesticides and generally affect only the target pest and closely related organisms. They fit well into IPM (Integrated Pest Management) programs for increased efficacy and yields and are useful for resistance management. In some countries biopesticide registration is less restrictive which allows for faster registrations than synthetic pesticides.

However, when it comes to biopesticides there are some challenges that need to be overcome. Generally biopesticides have a limited shelf life and can have limited persistence in the field. They can be slower acting meaning they are better for preventative measures as opposed to a curative method. Additionally they have to compete with conventional pesticides that are easy to use, low cost and generic.

Formulating with biopesticides

The goal when formulating biopesticides is to make these technologies more predictable and competitive than synthetic pesticides. When using a microbial active ingredient the formulation should: stabilise the organisms during production, distribution and storage, aid in the handling and application of the product, protect the microbe from harmful environmental factors and enhance the activity of the organism.

Biopesticides are living things or compounds derived from living things that deter pests (NPIC 2020). Many of these pest deterrents are naturally found in the environment. Biopesticides can be microbial organisms such as bacteria and fungi. One example is *Bacillus thuringiensis (Bt)*, which produces proteins that are toxic to immature insects or larvae.

Because these toxic proteins act specifically on certain larvae, they pose minimal risk to humans and other animals. Bacteria and fungi used as biopesticides are often target-specific, so they can be used to control specific weeds or insect pests. Plant-derived substances such as corn gluten, black pepper, and garlic compounds can be used as biopesticides to control insects.

Other types of biopesticides are naturally occurring insect hormones—which can repel bugs, disrupt their mating habits, or stunt their growth—and synthetic substances that have the same molecular formulas and use the same modes of action as their natural counterparts. Genetically modified plants are also considered forms of biopesticides because they have been engineered with pest-deterring genes and proteins from natural sources. For example, *Bt* genes have been transferred into some crops, which protects them from marauding insects without the need for additional inputs.

The EPA defines minimum risk pesticides as products that are exempt from the Federal Insecticide, Fungicide, and Rodenticide Act. These active and inert ingredients are listed under the minimum risk exemption regulations in the Code of Federal Regulations (CFR). The identification and standardization of minimum risks pesticides is enhanced by inclusion of Chemical Abstracts Service numbers for each (EPA 2015). To be categorized as a minimum risk pesticide, the following six conditions must be met (EPA 2019):

1. The product's active ingredients must be only those listed in 40 CFR 152.25(f)(1).
2. Any other ingredients, including inert ones, must also be on this list.
3. All of the ingredients, both active and inert, must be listed on the label.
4. The product cannot claim to control pests that affect human health.
5. The name of the producer and the company's contact information must be displayed prominently on the label.
6. The label cannot include any false or misleading statements.

Certain food-grade oils are also eligible to be minimum risk pesticide products. Castor, cinnamon, clove, corn, cottonseed, garlic, linseed, peppermint, rosemary, soybean, spearmint, and thyme oils fall into this category. Eligible food-grade spices are cinnamon, cloves, cornmint, peppermint, rosemary, spearmint, thyme, and white pepper. Citric acid, corn gluten meal, lauryl sulfate, potassium sorbate, putrescent whole egg solids, sesame, ground sesame plant, sodium chloride, and sodium laurel sulfate are also food-grade chemicals and by-products that pose minimum risks. Despite being natural, these products should still be considered toxic and used with care as many plant oils are harmful to the skin and eyes and some may cause severe reactions.

While many believe that organic foods are always pesticide-free, that is not true; organic crops are just as vulnerable to pest problems as conventional crops are. So when necessary, farmers rely on organic pesticides, which are usually made from naturally occurring substances but not always. The USDA's National List of Allowed and Prohibited Substances outlines which natural and synthetic substances are allowed as organic pesticides. Among this list are various soaps, hydrogen peroxide, sodium hypochlorite (chlorine bleach), diatomaceous earth, pyrethrins, and iron sulfate. Like conventional pesticides, organic pesticides can be quite toxic even if they are natural substances; for example, arsenic, strychnine, and nicotine are toxic natural compounds (these are not used as pesticides). Furthermore, organic pesticides can pose problems for bees, butterflies, and other crop pollinators and beneficial insects.

The Xerces Society is committed to protecting insect species that benefit crops and other plants and provides an overview of common organic pesticides with toxicity ratings and special precautions (Xerces Society 2019). The minimal risk pesticides food-grade oils and corn gluten have a low to medium impact on beneficial insects. Acetic acid concentrations above 10% are toxic to bees and other beneficial insects. The Xerces Society recommends applying organic pesticides at night to reduce exposure to beneficial insects, which are mostly active during the day. The organization also suggests low usage rates, rotating different substances, and combining low dosages for a hurdle effect. It is important to note that biopesticides are only minimally impactful if they are used thoughtfully.

According to Michelle Wright, an independent food regulation and food labeling consultant, of the three classifications of natural and biological pesticides listed on the NPIC's website, organic certification is the only route to making organic pesticide claims. Outside of organic certification, it is difficult to achieve the traceability that is needed for most alternative pesticide claims because commodity food ingredients are mixed during processing, storage, and transportation. In the future, consumers may demand more preharvest transparency and manufacturers may want to make alternative pesticide marketing claims. Therefore, preharvest supply chain traceability and accountability should continue to improve, mimicking advances in postharvest traceability.

There are many alternatives to conventional pesticides available to food producers and consumers. Helpful information on pesticide use and alternatives to conventional pesticides is readily available to counteract and supplant misinformation. Besides the NPIC's website, food producers and consumers can consult the USDA's NOP, the Organic Trade Association, and other web-based resources for sound information.

Integrated Pest Management of agricultural pests

Its most important goals:

Attempting to control pests at the lowest possible costs for the farmer, while minimizing environmental pollution whenever possible.

Consolidate the efforts being made to protect crops to give a positive solution to the problem Focusing on conducting integrated resistance operations to prevent the pest population from reaching the level of the economic limit of damage to allow time to start control operations before the density of the pest population reaches the economic limit of damage, and thus the control is effective and feasible.

To initiate an integrated control system, follow the following points:

1- Studying the types of pests that infect crops and their impact on crop productivity
2- The relationship of plant growth and its exposure to pests
3- The value of the yield of the crop
4- Natural enemies and environmental resistance that works against pests
5- The extent of the possibility of conducting different methods of resistance, chemical and non-chemical, and the costs of conducting them
6- The impact of different types of control on the environment and humans

Sound scientific foundations for the success of integrated Pest management programs

- Estimating the percentage of losses and determining when pest resistance will have a feasible economic return.
- Determine which control methods can be applied in integrated pest management programs.

The most successful methods in integrated pest management programs

1- Using plant varieties that are resistant to pest infestation.

2- Biological control.

3- Agricultural control.

4- Mechanical control.

5- Chemical control.

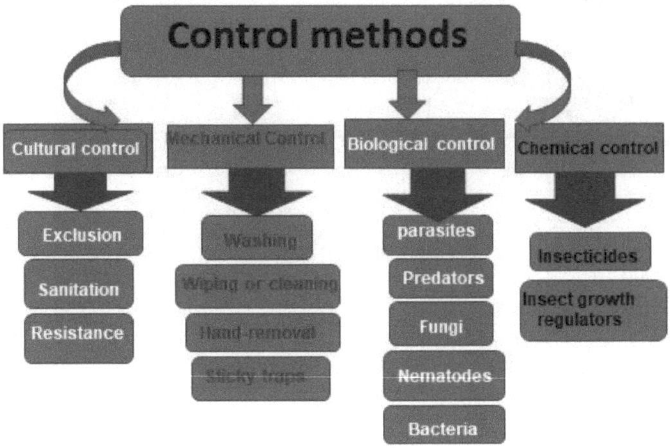

Sanitation
Use clean pots
Use only sterilized soil
Do not contaminate potting soil

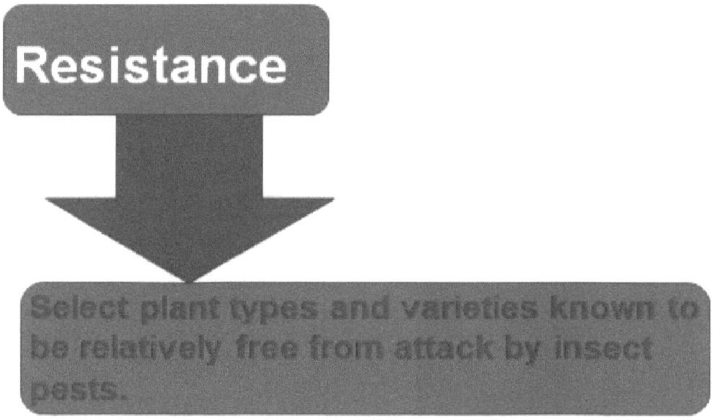
Resistance
Select plant types and varieties known to be relatively free from attack by insect pests.

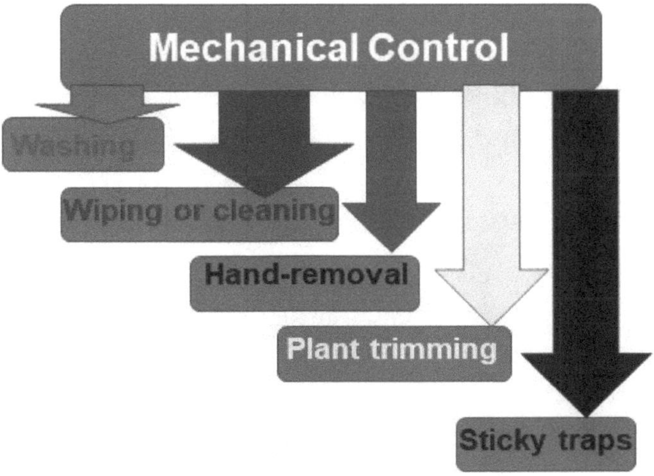
Mechanical Control
Washing
Wiping or cleaning
Hand-removal
Plant trimming
Sticky traps

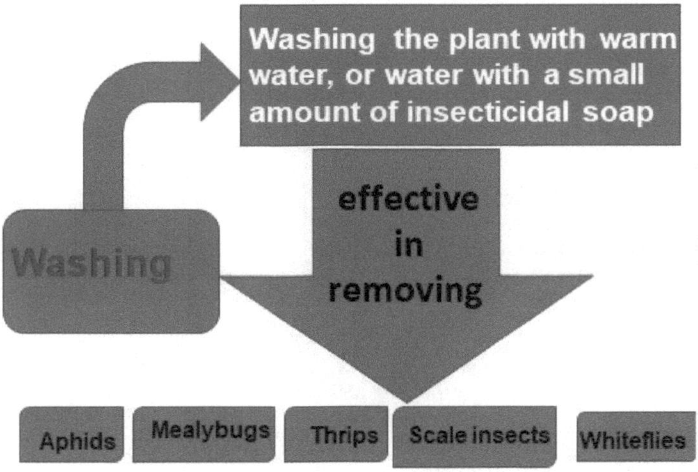
Washing the plant with warm water, or water with a small amount of insecticidal soap
Washing
effective in removing
Aphids
Mealybugs
Thrips
Scale insects
Whiteflies

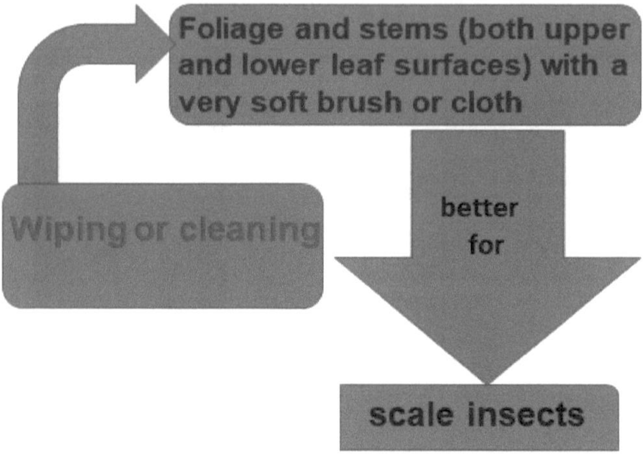
Foliage and stems (both upper and lower leaf surfaces) with a very soft brush or cloth
Wiping or cleaning
better for
scale insects

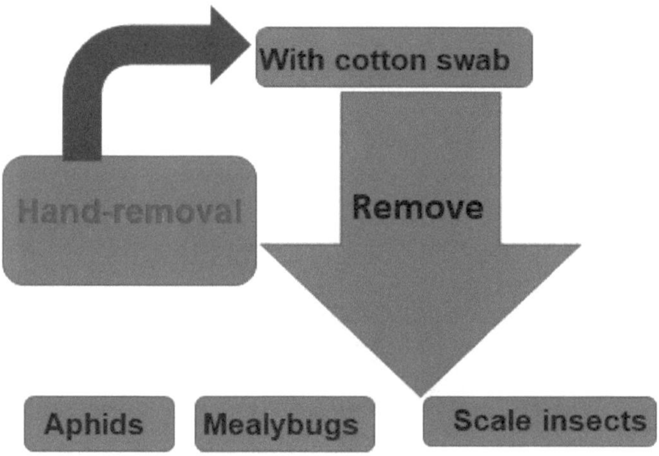
With cotton swab
Hand-removal
Remove
Aphids
Mealybugs
Scale insects

Sticky traps

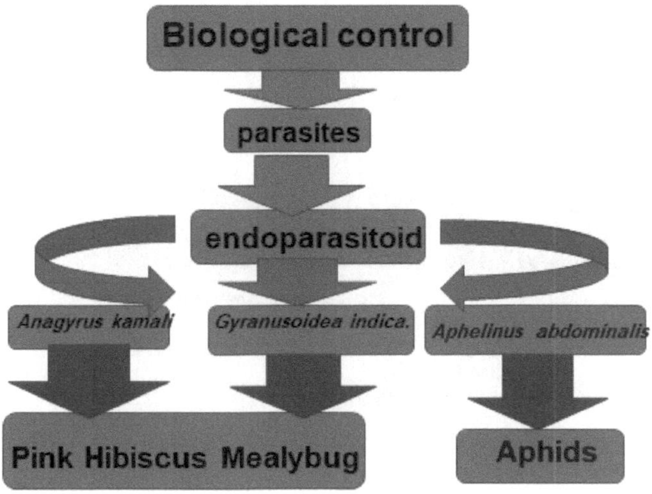

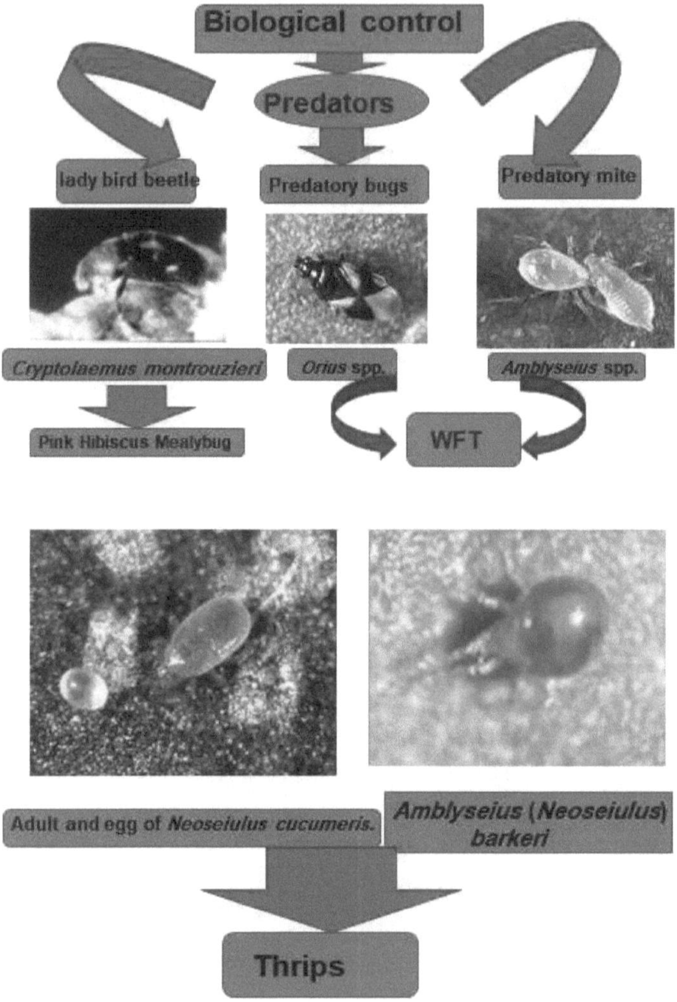

Biological control
Predators
lady bird beetle
Predatory bugs
Predatory mite
Cryptolaemus montrouzieri
Orius spp.
Amblyseius spp.
Pink Hibiscus Mealybug
WFT
Adult and egg of Neoseiulus cucumeris.
Amblyseius (Neoseiulus) barkeri
Thrips

Predacious mites (Fam. Phytoseudae)

Pest	Predatore
Mealy bug	*Euseius Scutalis*
Scale Insects	*Typhlodromus negevi*
	Typhlodromips swirski
Bemisia tabaci	*Euseius Scutalis*
	Typhlodromips swirski
Thrips	*Neoseiulus barkeri* (Hugles)
	Neoseiulus cucumerius

The following must be observed when using chemical control

- The use of pesticides with the importance of accurate timing when using them
- Protection of beneficial insects such as bees and pollinating insects to reach a high level of integrated control
- Determine the necessary dose
- Treatment method
- Use of specific pesticides
- Reducing the dose reduces the severity of contamination and the costs of resistance
- Use pesticides only when the pest has reached the critical economic limit
- Use the pesticide that causes the least harm to humans and animals

- Using the correct and safe control methods, by estimating the efficiency of all methods before using them. Therefore, the benefits and disadvantages of each method must be studied and an integrated plan for its use should be developed.

Biological control as one of the safe methods of pest control

What is a pest?

Every animal or plant has a numerical density limit that occurs It has economic damage.

Agricultural pests include insects pathogens (bacteria, fungi, viruses also from vertebrates such as rodents and birds

Insects appeared and spread widespread on the surface of the earth a long time ago far.

And from the moment a person began to think, in expanding the agricultural area occupied

The problem of protecting these crops from Damage to insects and other pests much of his thinking and effort.

Chemical control method acquired

I have a special interest in the aftermath of the war Second World.

Where was the genocidal property discovered?

For insects in the substance DDT was this as a new opening in the world of industry pesticides. The pesticide industry flourished Agricultural producers are about to use these pesticides.

The total reliance on pesticides resulted in Pest control leads to some drawbacks over time, it became clear during the sixties from this century, and then the awareness of this increased Negatives and their effects on the environment and humans. Hence the tendency to fight Vital to reduce pesticide use.

WARNING
PESTICIDES
FIRE WILL CAUSE
TOXIC FUMES

Biological or biological control

Biological or biological control is defined as the human use of a living organism in order to eliminate a specific pest.

(تريكوجراما) طفيل بيوض لدودة ثمار التفاح

Biological Enemy Items

Parasites: parasitize and feed on pests they are of two types, internal and external parasitism.

Predators: prey on and feed on pests.

Pathogens: They cause various types of diseases and kill it

Microbial biopesticides	Target pest/mechanism
A. radiobacter	Crown galls
Adoxophyes orana granulovirus (AoGVs)	Summer fruit tortrix moth (*Adoxophyes orana*)
Altemaria destruens	Cuscuta control
Ampelomyces quisqualis	Hyperparasitic, powdery mildew control and damping off disease control
Aspergillus flavus strain AF36	Fungicide for cotton Pistachio orchards
B. bassiana	Effective against variety of insects such as crickets, white grubs, fire ants, flea beetles, whiteflies, plant bugs, grasshoppers, thrips, aphids, mites, mosquito larvae and many others
B. firmus	Diamondback moths
B. popilliae	Stomach poison, larvae of various beetles

Microbial biopesticides	Target pest/mechanism
B. pumilus	Effective against rust, downy and powdery mildews
B. sphaericus	Stomach poison mosquitoes
B. subtilis	Effective against root rot caused by *Rhizoctonia*, *Fusarium*, *Alternaria*,*Aspergillus* and *Pythium*. Also effective against some foliar diseases
B. subtilis FZB24	Effective against *Rhizoctonia*, *Fusarium*, *Alternaria*, *Verticillium* and*Streptomyce*s on vegetables and ornamental plants
B. thuringiensis aizawai 7.29	Lepidoptera
B. thuringiensis aizawai IC 1	*Lepidoptera*, *Diptera*
B. thuringiensis Berliner	*Lepidoptera*
B. thuringiensisentomocidus 6.01	*Lepidoptera*
B. thuringiensis israelensis	*Diptera*
B. thuringiensis japonensis Strain Buibui	Activity against soil-inhabiting beetles from cotton
B. thuringiensis kurstaki HD-1	*Lepidoptera, Diptera*
B. thuringiensiskurstaki KTO,	*Lepidoptera*

Microbial biopesticides	Target pest/mechanism
HD-1	
B. thuringiensis morrisoni PG14	*Diptera*
B. thuringiensis ssp. Kurstakiandaizawai	Act as insecticides against lepidopteran larval species from Soyabean castor, tobacco
B. thuringiensis tenebrionis	Activity against coleopteran adults and larvae, most notably the Colorado potato beetle (*Leptinotarsa decemlineata*) From tomato, Chilies, rice red gram,
B. thuringiensis tenebrionis (san diego)	*Coleoptera*
B. thuringiensis var. aizawai	Stomach poison, effective against lepidopterans in vegetables and maize
B. thuringiensis var. galleriae	Stomach poison, effective against American bollworm on cotton, tobacco caterpillar on chillies, leaf folder of rice and diamondback moth on vegetables
B. thuringiensis var. israelensis	Stomach poison, effective against mosquitoes and black fly larvae and midges
B. thuringiensis var. kurstaki	Stomach poison, effective against most lepidopteran larvae and some leaf beetles
B. thuringiensis var. tenebrionis	Stomach poison, effective against coleopteran beetles on vegetables

Microbial biopesticides	Target pest/mechanism
Bacillus firmus	Nematodes
Bacillus sphaericus	Diamondback moths
Bacillus subtilis	Pathogens such as *Rhizoctonia, Fusarium, Aspergillus*, banana Sigatoka (caused by *Mycosphaerella musicola*), Paddy (bacterial leaf blight)
Bacillus thuringiensis aizawai	Armyworms, diamondback moth
Bacillus thuringiensis and *Bacillus sphaericus 2362 (Bs)*	Effective against mosquito and other dipteran larvae black fly (simuliidae), and fungus gnats from cabbage and Cauliflower
Bacillus thuringiensis israelensis	Mosquitoes and black flies
Bacillus thuringiensis kurstaki	Lepidoptera
Bacillus thuringiensis sphaericus	Mosquitoes
Bacillus thuringiensis subsp. Israelensis	Lepidopteran pests
Bacillus thuringiensis subsp. Kurstaki	Lepidopteran pests
Bacillus thuringiensis tenebrionis	Colorado potato beetle
Bacillus thuringiensis var.	*Helicoverpa armigera*

Microbial biopesticides	**Target pest/mechanism**
galleriae	
Bacillus thuringiensis var. israelensis	Diamondback moths
Bacillus thuringiensis var. kurstaki	Diamondback moths
Beauveria bassiana	Mango hoppers and mealy bugs and coffee pod borer
Beauveria bassiana	Wide range of insects and mites
Beauveria bassiana	Many insect species such as aphids, whiteflies, thrips
Beauveria brongniartii	*Helicoverpa armigera*, berry borer, root grubs
Burkholderia cepacia	Effective against soil fungal pathogens
Burkholderia spp.	Chewing and sucking insects and mites; nematodes
Candida oleophila	Effective against postharvest pathogens like botrytis and Penicillium
Chromobacterium subtsugae	Chewing and sucking insects and mites
Close analog of the fumonisin mycotoxins	Highly toxic used to kill weed species. Highly toxic to mammals by inhibition of ceramide synthase. Broadleaf plants (e.g. jimsonweed, prickly sida and black

Microbial biopesticides	Target pest/mechanism
	nightshade)
Coniothyrium minitans	Effective against Sclerotinia species on canola, sunflower, peanut, soyabean and vegetables
Cryptolaemusm ontrouzieri	To control mealy bugs especially in fruits like citrus, coffee, grapes and several other fruit crops and ornamental
Cryptophlebia leucotreta granulovirus	False codling moth (*Thaumatotibia leucotreta*)
Cydia pomonella granulovirus (CpGVs)	*Cydia pomonella*
Entomopathogenic nematodes such as Steinernema carpocapsae	Thrips, fungus gnats (*Bradysia* spp.), mole crickets (*Scapteriscus* spp.) and other insects
Erwinia amylovora (HrpN harpin protein)	Insecticide, fungicide and nematicide. Multi-spectrum
Fusarium oxysporum (non-pathogenic)	Effective against pathogenic *Fusarium* on basil, carnation, cyclamen, tomato
Gliocladium catenulatum Gliocladium spp.	Effective against *Pythium, Rhizoctona, Botrytis* and *Didymella* species on greenhouse crops
Gliocladium virens	Effective against soil pathogens causing damping off and root rot

Microbial biopesticides	Target pest/mechanism
Granulosis virus	Effective against leafroller and codling moth and other lepidopterans
H. bacteriophora	Effective against many lepidopteran larvae, turf and Japanese beetles and soil insects
H. bacteriophora	Borers
H. megidis	Effective against black vine weevils and soil insects
Helicoverpa armigera NPV	*H. armigera*
Helicoverpa armigera Nucleopolyhedrosis virus and (nucleopolyhedrosis) NPV for other *Lepidoptera*	*Helicoverpa* spp. (for HNPV) and other *Lepidoptera* species
Helicoverpa armigera nucleopolyhedrovirus (HearNPV)	African cotton bollworm (*Helicoverpa armigera*), corn earworm (*H. zea*) and other *Helicoverpa* species (*H. virescens, H. punctigera*)
Helicoverpa zea Nuclear Polyhedrosis Virus	*Heliothis* and *Helicoverpa* species
Helicoverpa zea nucleopolyhedrovirus	*Helicoverpa* spp. and *Heliothis virescens*
Helicoverpaarmigera (Hubner) and*Spodopteralitura (Fabricius)*	Used against bollworms in cotton and pod borers species specific for*Lepidoptera*, *Hymenoptera* and *Diptera* fr

Microbial biopesticides	**Target pest/mechanism**
	om cotton, flax, groundnuts, jute, maize, soya beans, tea, tobacco, vegetables, lucerne
Heterorhabditis bacteriophora	*Otiorhynchus* spp., chestnut, moths, black vine weevil and soil-dwelling beetle larvae, *Melolontha*, caterpillars, cutworms, leafminers
Heterorhabditis downesi	Black Vine weevil, *Otiorhynchus sulcatus*
Hirsutella thompsonii	Stimulates premature fungal epizootics, effective against mites
Hirsutella thompsonii *Isaria fumosorosea*	Spider mites
Lagenidium giganteum	Effective against mosquito larvae and related dipterans. Kill through zoospores
Lecanicillium lecanii	Aphids, leafminers, mealybugs, scale insects, thrips, whiteflies
Lymantria dispar multiple nucleopolyhedrovirus (LdMNPV)	*Lymantria dispar*
M. anisopliae	Effective against range of pests. Green muscle is specific for locusts and grasshoppers
Metarhizium anisopliae	Beetles & caterpillar pests; grasshoppers, termites

Microbial biopesticides	Target pest/mechanism
Metarhizium anisopliae	Insects such as crickets and grasshoppers
Metarhizium anisopliae *Entomopathogenic* fungus	Insect pest control used as a biocontrol agent, particularly for malaria vector species from cabbage, canola (*Brassica napus* L.)
Myrothecium vaerrcaria *Paecilomyces fumosoroseus*	Black vine weevil, effective against many nematodes effective against whiteflies in greenhouse
Myrothecium verrucaria	Nematodes
Neodiprion abietis nucleopolyhedrovirus (NeabNPV)	*Neodiprion abietis*
NPV for *A. gemmatalis*	Effective against velvetbean caterpillar and sugarcane borer
NPV for *Anagrapha falcifera*	Indian meal moth, effective against lepidopterans
NPV for *Autographa calofornica*	Effective against Alfalfa looper
NPV for *H. zea and H. virescens*	Effective against bollworms
NPV for *L. dispar*	Effective against gypsy moth
NPV for *Mamestra brassicae*	Effective against lepidopterans
NPV for *Neodiprion sertifer,*	Effective against sawfly larvae

Microbial biopesticides	**Target pest/mechanism**
N. lecontei and N. abietis	
NPV for *Orgyia pseudotsugata*	Effective against Douglas-fir tussock moth
NPV for *Spodoptera exigua*	Effective against beet and lesser army worms, pig weed caterpillar and mottled willow moth
NPV for *Syngrapha falcif*	Effective against *Helicoverpa* and *Cydia* spp.
NPV of Spodoptera litura	*Spodoptera litura*
P. hermaphrodita	Effective against slugs eating nematode
Paecilomyces fumosoroseus	Insects, mites, nematodes, thrips
Paecilomyces lilacinus	Effective against nematodes
Paecilomyces lilacinus	Plant pathogenic nematodes
Phasmarhabditis hermaphrodita	Molluscs
Phelbia gigantean	Management of fungal pathogens in various crops
Phromone lures for*Helicoverpaarmigera (Hubner)* and*Spodopteralitura*	To trap productive male of gram borer and tobacco caterpillar,*Spodopteralitura* from tomato, potato: Cabbage: *Brassica oleracea*, brinjal
Plutella xylostella granulovirus	*Plutella xylostella*

Microbial biopesticides	Target pest/mechanism
Pseudomonas fluorescens	Whitefly
Pseudomonas fluorescens,	Several fungal, viral, and bacterial diseases such as frost forming bacteria. Wheat (Loose Smut) Paddy (bacterial leaf blight) Ground nut (late leaf spot) rice (leaf and neck blast) chili seedlings (damping off) tomato (wilt)
S. carpocapsae	Effective against black vine weevils, strawberry root weevils, cutworms, cranberry girdler and termites
S. glaseri	Effective against root weevils, cutworms, fleas, borers and fungal gnats
S. viridochromogenes *Streptomyces hygroscopicus*	Anntnual and perennial broadleaf weeds and grasses. L-phosphinothricin inhibitor of glutamine synthetase (GS)
Saccharopolyspora spinosa *(Actinomycete* bacteria)	Used for the control of a very wide range of caterpillars, leaf miners, thrips and foliage feeding beetles
Saccharopolyspora spinosa	Insects
Serratia entomophila	Grass grub
Spodoptera exigua nucleopolyhedrovirus (SeNPVs)	*Spodoptera exigua*

Microbial biopesticides	**Target pest/mechanism**
Spodoptera littoralis nucleopolyhedrovirus (SpliNPV)	African cotton leaf worm (*Spodoptera littoralis*)
Spodoptera litura NPV	*S. litura*
Spodoptera litura nucleopolyhedrovirus	*Spodoptera litura*
Steinernema carpocapsae	Borer beetles, caterpillars, cranefly, moth larvae, *Rhynchophorus ferrugineus*, Tipulidae.
Steinernema feltiae	Bradysia spp., *Chromatomyia syngenesiae*, *Phytomyza vitalbae*, soil dwelling pests, codling moth larvae, sciarids, thrips
Steinernema kraussei	Vine weevil larvae
Streptomyces avermitilis (soil bacterium)	Ornamental plants, citrus, cotton, pears and vegetable crops to control phytophagous mites
Streptomyces griseoviridis	Effective against wilt, seed rot and stem rot
Streptomyces hygroscopicus	Morning glories, hemp (*Sesbaniabispinosa*), Pennsylvania smartweed (*Polygonum pensylvanicum*) and yellow nutsedge. Natural GS inhibitors, used for weed control in Japan.

Microbial biopesticides	**Target pest/mechanism**
Streptomyces lydicus	Effective against soil borne diseases of turf, nursery crops
Streptomyces sp. culture broths	Useful in glutamate and GABA gated chloride-channel opening. From fruit, vegetables and ornamentals plants
T. harzianum + T. viride	Effective against Armillaria and botryoshaeria and others
Talaromyces flavus V117b	Effective against fungal pathogens of tomato, cucumber, strawberry and rape oilseeds
Trichoderma viride/harzanum	Control root rot and wilt diseases especially on pulses and Paddy (Foot rot)
Trichoderma harzianum	Effective against variety of soil pathogens and wound pathogens
Trichoderma sp.	Suppresses root pathogens
Trichoderma sp. (egg parasite)	Used for control of sugarcane, early shoot borer, cotton bollworms, sorghum stem borer
Trichoderma viride	Effective against rot diseases
Trichoderma viride	Root rots and wilts
Trichogramma parasitoid	Sugarcane borers
Trichogramma spp.	Eggs of Lepidoptera
Verticillium lecanii	Mealy bugs and sucking insects

The entomopathogenic fungi

- There are fungus species that can infect and kill insects.
- These fungi are called entomopathogenic fungi or insect pathogenic fungi.
- These fungi are very specific to insects, often to particular insect species, and do not infect animals or plants.
- Most insect- pathogenic fungi need humid conditions for infection and development but some fungus species can also infect insects when it is dry.
- When an insect has been killed by a fungus, the fungus emerges from the insects body to produce spores that can sometimes be seen as a "dusty" or "cottony" appearance.
- When you find many insects dead and covered with fungus, you can consider making your own bio-insecticide from these dead insects.
- Collect as many as you can find in the field, put them in a jar with water, crush them a little and stir well.
- This will release fungus spores into the water.
- Filter the water slightly to remove large insect parts.
- The remaining solution can be used to test.
- its Spray the solution over insects that are placed in a jar.
- Check if these insects become infected over the next days.
- Spray another jar of insects with pure water as a control.

Classification of entomopathogenic fungi

The entomopathogenic fungi are found in the following divisions: Zygomycota, Ascomycota Deuteromycota, and also in Chytridiomycota and Oomycota, which were formerly classified as fungi.

Several genera of entomopathogenic fungi are currently under investigation, either belonging to the order Entomophthorales of the division Zygomycota or to the phylum Hyphomycetes of the division Deuteromycota.

Most, if not all, entomopathogenic fungi have life cycles that coincide with insect host stages and environmental conditions. With differences between Hyphomycetes and Zygomycetes, especially Entomophthorales.

Fungal infection transmission

Asexually produced fungal spores (conidia) are usually responsible for infection and are spread throughout the environment in which the insect host is present. When conidia settle on the cuticle of a suitable host, where they attach and grow a series of reactions and activation of enzymes by the host and the fungal parasite begin.

Invasion of the body and blood circulation of insects occurs once when the fungus passes through the cuticle of the insects' exoskeleton. Insect tissue invasion processes are similar to plant pathogens, including germ tube formation, adhesion, and peg penetration.

Mode of action of entomopathogenic Fungi

With most Fungi disease development can be divided into nine steps as given by Roberts (1980)

1-Attachment of the infective unit.

2- Germination.

3-Penetration

4-Multiplication

5-Production of toxic metabolites.

6- Death of the host.

7-Growth in the mycelial phase with invasion of virtually all host organ

8-Penetration of hyphae from the interior through the cuticle to the exterior of the insect.

9-Production of new infection units (conidia).

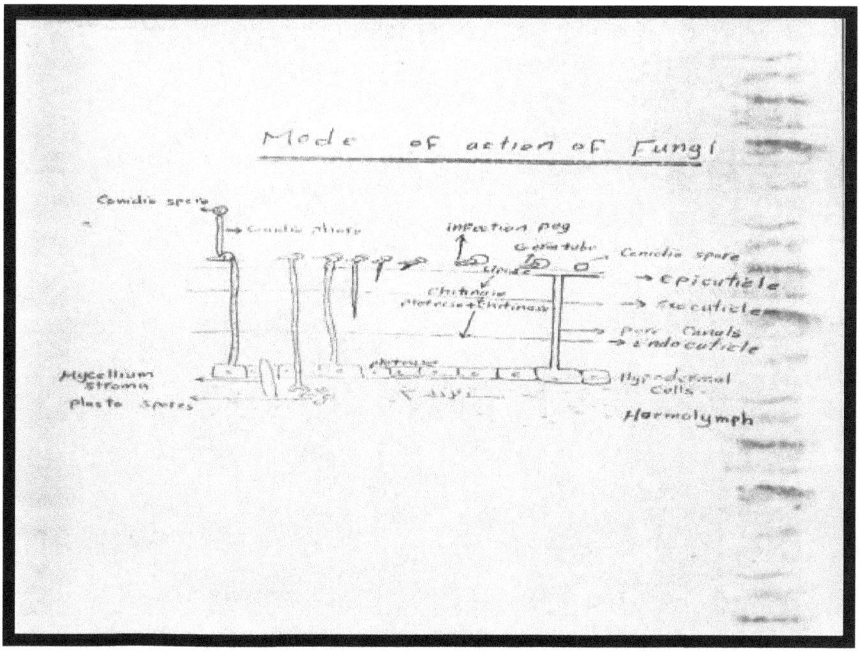
Mode of action of Fungi
Conidia spore
Infection peg
Germ tube
Conidia spore
epicuticle
Chitinase
Protease+Chitinase
Pore Canals
Endocuticle
Mycelium
Stroma
Blasto spores
Cells
Hermolymph

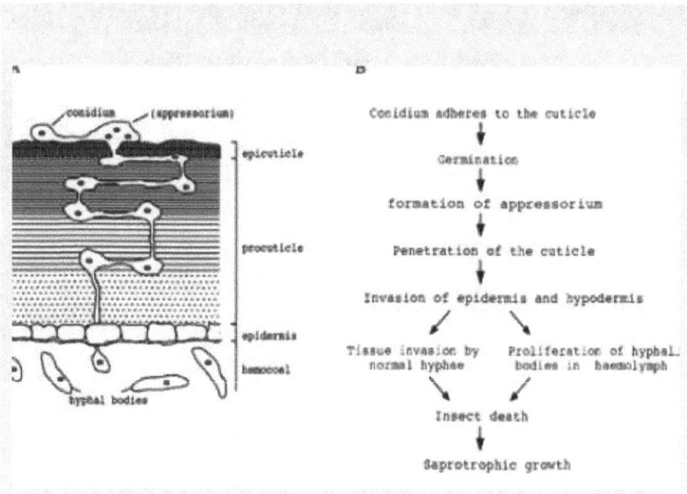
A
conidium
(appressorium)
epicuticle
procuticle
epidermis
hemocoel
hyphal bodies
B
Conidium adheres to the cuticle
Germination
formation of appressorium
Penetration of the cuticle
Invasion of epidermis and hypodermis
Tissue invasion by normal hyphae
Proliferation of hyphal bodies in haemolymph
Insect death
Saprotrophic growth

Prof. Mohamed Abdel-Raheem, 2022

B. bassiana

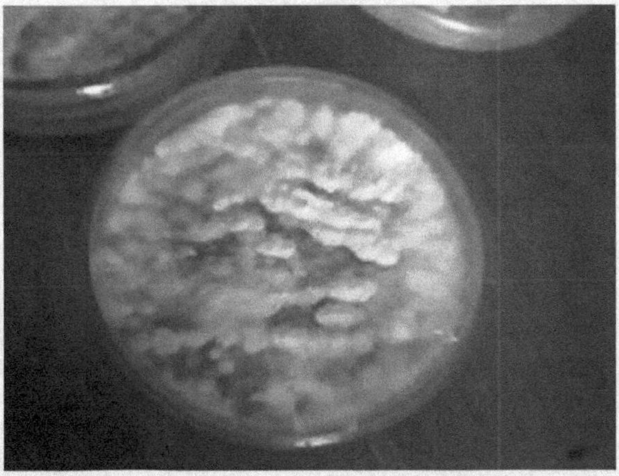

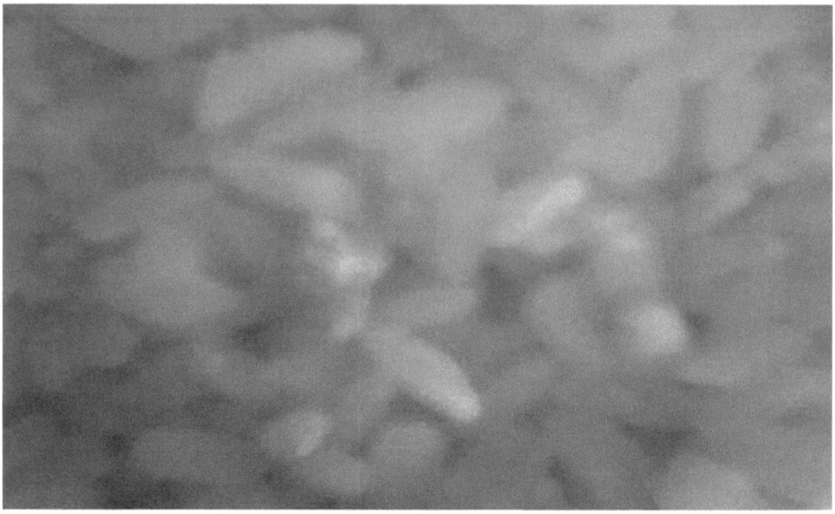

***Beauveria bassiana*:**

- This fungus is found naturally on some plants and in the soil.
- It needs warm, humid weather for spread and infection.
- Infected insect larvae eventually turn white or gray.
- It can infect larvae of rice insects like black bugs and rice seed bugs but also pests of many other crops like maize, potato, and vegetables.
- It is commercially available from many companies in the U.S. and Europe.
- Unfortunately, some natural enemies such as lady beetles can be susceptible to *Beauveria*.
- One possible application method that may avoid harming beneficial insects is the use of fungus- contaminated food baits that are only attractive to pest species.

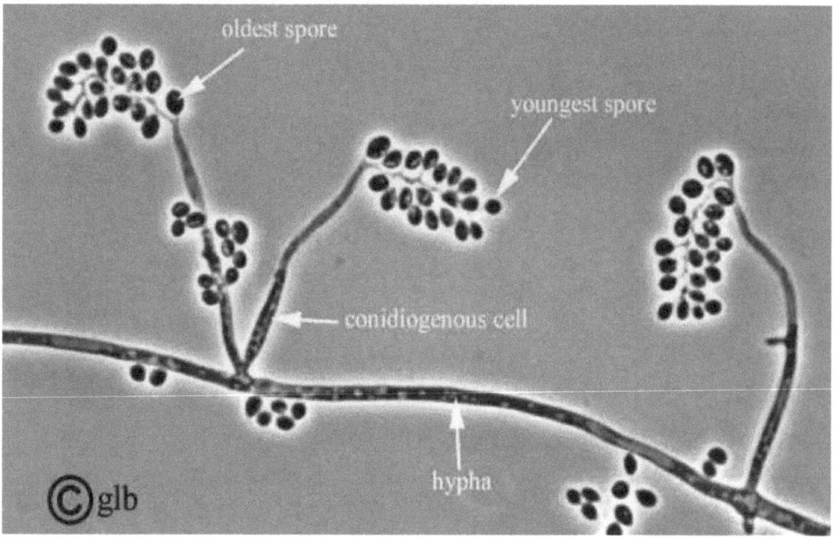

***Metarhizium* species**

- *Metarhizium* species is being tested as natural enemy of many soil insect pests.
- *Metarhizium* can be an important control agent of aphids.
- In Indonesia (West Sumatra), *Metarhizium* is used as a spraying solution for control of different pest insects in chili.
- In Philippines, *Metarhizium* effectively reduced populations of rhinoceros beetles in coconut (FAO Dalat report, 1998).
- In the United States *Metarhizium* is commercially available as a foliar spray.

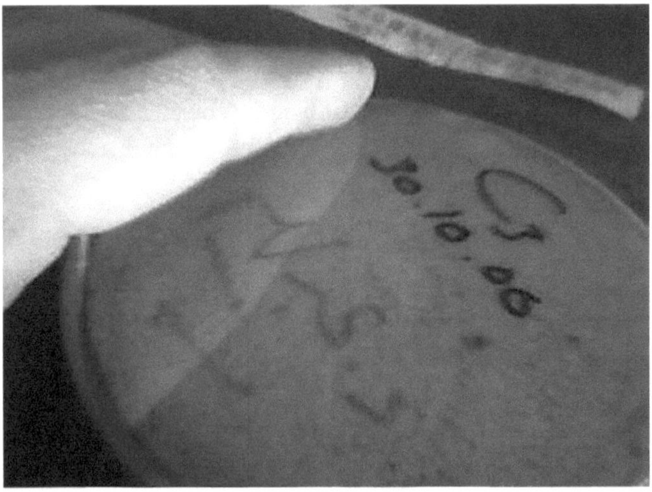

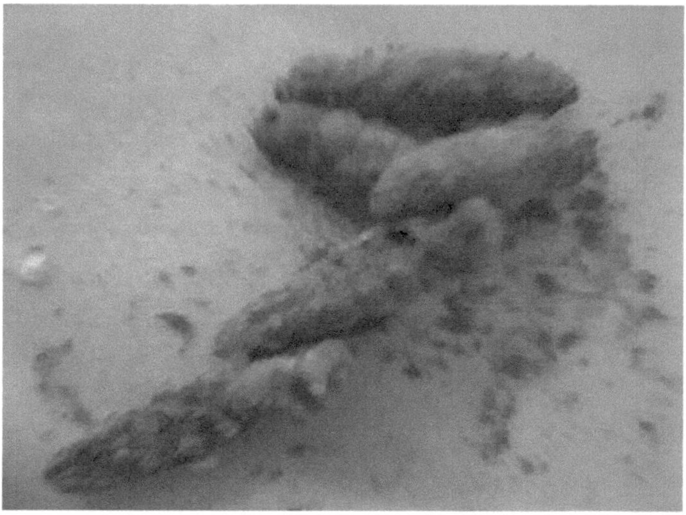

***Entomophthora sp.*:**

- This fungus is fairly specific to aphids and several species of flies.
- In Indonesia, an *Entomophthora* has been found on a cabbage caterpillar.
- However, methods for production and distribution for these fungi have not been developed, so they are not available commercially (FAO Dalat report, 1998).

Nomuraea rileyi

- *Nomuraea rileyi* occurs naturally in Indonesia as a natural enemy of a cabbage caterpillar (FAO Dalat report, 1998).
- *Nomuraea* often causes large disease outbreaks in caterpillars in soybean.
- Methods for production and distribution of this fungus have not been developed, so it is not available commercially.

Paecilomyces

- A species of *Paecilomyces* was found infecting whiteflies on cabbage in Cebu, Philippines.
- Researchers from the Regional Crop Protection Center have isolated the species and are now mass-producing it on artificial media (FAO Dalat report, 1998).
- In other countries, for example USA, *Paecilomyces fumosoroseus* is commercially available for whitefly control.
- It can be applied as a spore solution and it has some activity against aphids, thrips and spider mites.

Verticillium sp.

- *Verticillium* sp. is used in Europe against whitefly, thrips and aphids, especially in greenhouse crops where controlled environment favors fungus effectiveness.
- *Verticillium lecanii* is commercially available in Europe and USA for the control of greenhouse whitefly.

Zoophthora radicans

- *Zoophthora radicans* has a wide host range.
- It has been found infecting cabbage aphids and caterpillars.
- This fungus forms a white to tan colored mat of hypae (fungus threads) that covers the insect completely.
  Numerous spores are formed on this mat.
- *Zoophthora radicans* is found in so many insects, it is suspected that this may actually be several fungus species, each able to attack only a few insect species (Shepard *et al*, 1999).

How to protect and help fungi

- Many insect-pathogenic fungi live in the soil.
- There is evidence that application of some soil insecticides, fungicides and herbicides can inhibit or kill these fungi.
- For example, even very low concentrations of some herbicides can severely limit the germination and growth of *Beauveria bassiana* fungal spores in soil samples.

Primary or secondary • ???

- When you leave a plate of food with for example some chicken meat in your kitchen and leave it for a few days, you may find the meat covered with fungus when you look again.
- This fungus is called a secondary infection:
- It was not the reason the chicken died (the chicken probably died because a person killed it), and the fungus started growing after the chicken was dead.
- Similarly, when insects are dead, some fungi may start growing on the dead insect.
- This is also called a secondary infection.

- These fungi that cause secondary infection are part of the “trash men team” of nature:
- They make sure dead things are decomposed quickly.
- When the fungi actually cause a living insect to die, like insect-pathogenic fungi do
- It is called a primary infection

Interesting, but who cares?

- When searching for beneficial fungi in the field, it is important to distinguish between fungi causing primary infection (pathogens) and those causing a secondary infection.
- Fungi that cause secondary infections are not really pathogens (will not infect and Kill healthy insects).

Bacillus thuringiensis (Bt).

Bacillus thuringiensis (Bt) is a species of bacteria that lives in soil. It makes proteins that are toxic to some insects when eaten, but not others. The proteins are not toxic to humans because, like all mammals, we cannot activate them. *Bt* is not toxic to non-target wildlife. However, one type of *Bt* (aizawi) can be toxic to honeybees. *Bt* is used as an **insecticide**, typically, for insect larvae. Remember, it has to be eaten to work.

Bacillus thuringiensis (Bt) is a soil bacterium that forms spores during the stationary phase of its growth cycle. The spores contain crystals, predominantly comprising one or more Cry and/or Cyt proteins (also known as δ-endotoxins) that have potent and specific insecticidal activity.

Different strains of Bt produce different types of toxin, each of which affects a narrow taxonomic group of insects. Therefore, Bt toxins have been used as topical pesticides to protect crops, and more recently the proteins have been expressed in transgenic plants to confer inherent pest resistance.

Bt transgenic crops have been overwhelmingly successful and beneficial, leading to higher yields and reducing the use of chemical pesticides and fossil fuels. However, their deployment has attracted some criticism particularly with regard to the potential evolution of pest-resistant insect strains. Here, we review recent progress in the development of Bt technology and the countermeasures that have been introduced to prevent the evolution of resistant insect populations.

The insecticidal properties of Bt were recognized many years before the bacterium was identified, with some accounts suggesting that Bt spores may have already been in use in ancient Egypt. In the modern era, the bacterium was isolated in 1901 by the Japanese biologist Shigetane Ishiwatari during an investigation into wilt disease in silk worms, and he named it *Bacillus sotto*.

Ten years later, the same bacterium was isolated by Ernst Berliner from a diseased Mediterranean flour moth (*Ephestia kuehniella*) in the German province of Thuringia, and it was named *Bacillus thuringiensis* (**Siegel, 2000**).

The defining feature of Bt is its ability to produce proteinaceous crystals during sporulation. Bt is a member of the*Bacillus cereus* group of Gram-positive, spore-forming soil bacteria, and occasionally the bacteria lose their ability to form crystals and then become indistinguishable from *B. cereus*itself.

Similarly, *B. cereus* can be transformed into Bt, and investigations into the transformation mechanism led to the discovery that crystal formation is conferred by genes carried on a plasmid. The genes, which encode Cry/Cyt proteins, become active during sporulation because they are controlled by a dedicated RNA polymerase that is also synthesized specifically while spores are forming. Up to 20% of the spore protein content is represented by these Cry/Cyt toxins (**Aronson, 2002**).

The insecticidal properties of the crystals were discovered when dead flour moth caterpillars were found to be loaded with spores and crystals. Direct contact between the spores/crystals and healthy caterpillars had no effect, but when the spores and crystals were coated onto leaves, the caterpillars stopped feeding and died. After recognizing the potential of Bt as an insecticide, **Mattes (1927)** isolated the Bt strain discovered by Ernst and subsequent field trials against the European corn borer (*Ostrinia nubilalis*) gave promising results (**Husz, 1930**).

This work eventually led to the development of Sporeine, a commercial Bt insecticide, which was used for the first time in 1938.

Sporeine targeted lepidopteran insects, and the active ingredient was initially thought to be a food-borne invasive pathogen. However, this theory was discarded when direct injection of crystals into caterpillars was shown to have no effect.

Careful time-course analysis of the guts of caterpillars feeding on contaminated food revealed the disruption of cilia on the brush border of midgut epithelial cells shortly after ingestion of the crystals, followed by cell swelling and lysis.

The gut contents (including bacterial spores) were thus released into the body cavity, allowing bacteria to breed. Once the supply of nutrients was exhausted, the bacteria formed spores that could spread to a new host feeding on either the cadaver or the contaminated vegetation.

These early experiments showed that Bt toxins needed to be activated in the gut, and it was soon discovered that the critical factors were an alkaline environment and the presence of specific proteases, which cleaved the innocuous pro-toxin into its active form.

Once activated by proteolysis, each toxin binds to receptors in the brush border membrane and causes pores to open, disrupting the movement of solutes across the gut epithelium and causing the influx of water. The toxins were shown to be orally lethal to caterpillars in pure form, and the pro-toxins could be converted into active toxins *in vitro*, using specific proteases under alkaline conditions. The requirement for alkaline conditions, specific proteases and specific receptors explains why Bt is harmless to mammals (which have an acidic gut and lack the corresponding receptors) and why each toxin has a narrow host range.

Toxin structure and specificity

Many researchers have attempted to introduce taxonomic classification systems for Bt, using various criteria such as serotyping, phage susceptibility and plasmid profiles, and this has resulted in the classification of approximately 100 subspecies.

Although there is a good correlation between Bt subspecies and insect host range at the family level, the relationship tends to break down at the genus and species levels because most Bt strains can synthesize more than one toxin, resulting in complex and overlapping host profiles.

For example, most Bt *kurstaki* strains are specific for lepidopteran insects (butterflies and moths), whereas *israelensis* strains are specific for dipterans (flies) and *morrisoni* strains are specific for coleopterans (beetles). Other strains are not active against insects at all, but are toxic towards different invertebrates. For example, Bt strains containing only Cry5- and Cry6-type toxins are active against nematodes.

At the genus and species level, it is more useful to classify Bt strains functionally on the basis of which toxin proteins they produce, as this is a more logical way to define the host range. The toxins can be described in terms of their amino acid sequences, protein structures and modes of activity (**Crickmore *et al.*, 1998**).

Cry toxins interact with specific receptors located on the surface of midgut epithelial cells and are activated by host proteases following receptor binding, resulting in the formation of a pre-pore oligomeric structure that is insertion competent. In contrast, Cyt toxins directly interact with membrane lipids and insert into the membrane. The known Cry and Cyt proteins now fall into 32 sets including Cyt1, Cyt2 and Cry1 to Cry67 (**Crickmore *et al.*, 2010**).

Open in figure viewerPowerPoint

Phylogenetic trees representing (a) three-domain Cry proteins, and (b) related proteins (Cyt, Bin and Mtx).

Source: **http://www.lifesci.sussex.ac.uk/home/Neil_Crickmore/Bt/**.

The phylogenetic trees are modified from a TREEVIEW visualization of NEIGHBOR treatment of a CLUSTALW multiple alignment and distance matrix of the full-length toxin sequences, as described by **Crickmore *et al.* (2010)**. The grey vertical bars demarcate the four nomenclature ranks.

Despite their sequence diversity, all Cry proteins share a similar overall tertiary structure, as exemplified by the six structures solved thus far by X-ray crystallography (Cry1Aa, Cry2Aa, Cry3Aa, Cry3Bb, Cry4Aa and Cry4Ba).

The C-terminal portion is involved in crystal formation but is not part of the mature toxin, as it is cleaved off in the insect gut. The N-terminal portion is the toxin itself, and it comprises three domains. Domain I is a bundle of seven α-helices, six of which are amphipathic encircling the seventh hydrophobic helix, and this domain is responsible for membrane insertion and pore formation.

Domain II consists of three anti-parallel β-sheets with exposed loop regions, and domain III is a β-sandwich. Both domains confer receptor binding specificity thus helping to define the host range (**Boonserm *et al.*, 2006**).

A current model suggests that domains II and III initially bind to primary receptors (cadherins) which cleave the toxin within domain I and induce oligomerization, which in turn promotes binding to high-affinity secondary receptors tethered to the membrane via C-terminal glycosylphosphatidylinositol anchors (**Soberón *et al.*, 2009**).

The requirement for oligomerization has recently been confirmed through the isolation of dominant-negative mutations of Cry1Ab (**Rodríguez-Almazán *et al.*, 2009**).

An alternative model (**Zhang *et al.*, 2006**) suggests that initial binding triggers a Mg^{2+}-dependent signalling cascade that causes G-protein dependent cAMP accumulation and the activation of protein kinase A. Phylogenetic analysis has established that the diversity of the Cry family evolved by the independent evolution of the three domains and by swapping of domain III among toxins.

Development of Bt topical pesticides

Although Sporeine was used from 1938 in France, it was not registered as a pesticide in the United States until 1961. By this time, other Bt products such as Thuricide were already on the market, most based on *kurstaki* strain HD1 and active against lepidopteran pests (**Baum*et al.*, 1999**).

Despite their beneficial properties, these early Bt products did not compete well against chemical pesticides because of their poor performance. Commercial activities concentrated on two strategies to overcome these challenges, namely process development to increase the efficiency of the Bt products, and strain improvement to increase the intrinsic toxicity of the bacteria.

Process development began with improved fermentation and harvesting procedures, but perhaps the most important aspect was formulation. Liquid suspensions were the most convenient to handle, but tended to deteriorate in storage, while powders were easier to store and transport but drying was expensive and reformulation more complicated for the end user.

These challenges were addressed by the addition of excipients such as suspension agents to prevent suspensions settling and preservatives to increase shelf life, and in the case of powders by adding chemicals to improve pouring and wetting. UV screening agents were also used, to prevent rapid photolysis after spraying (**Burges and Jones, 1998a**).

These improvements, in addition to more rigorous quality control and the standardization of potency testing, led to a sixfold or more increase in efficacy in the field (**Burges and Jones, 1998b**).

Strain improvement in the 1960s led to the replacement of many of the early products with new Bt strains that were up to 10-fold more potent than their predecessors, and the search for new and better strains continues to this day.

Most Bt products are derived from *kurstaki*HD1 (e.g. Biobit, Dipel and Thuricide) although other strains are used to tackle lepidopteran pests (*kurstaki* SA-11, *kurstaki* SA-12), diptera (*israeliensis*) and coleoptera (*tenebrionis*) (**Kaur, 2000**). A selection of Bt topical pesticides is listed

Strain search and assessment programs initially involved bioassays and biochemical testing, but this has been replaced by PCR testing for specific toxin signatures. This can determine whether increased potency is achieved by higher toxin expression levels or the presence of a novel toxin, the latter offering the prospect of controlling different ranges of pests (**Kuo and Chak, 1996**; **Porcar and Juárez-Pérez, 2003**).

Tailoring the host range can be achieved not only through the discovery of new toxins, but also by creating new bacterial strains carrying previously unknown combinations of existing toxins, a process that can be implemented by conjugation or direct transformation (**González *et al.*, 1982**; **Kronstad *et al.*, 1983**; **Carlton and Gonzalez, 1985**).

Examples of combination pesticides produced by conjugation include Foil, a product based on strain EG2424, which produces Cry1Ac from Bt *kurstaki* (active against the European corn borer), and Cry3A from Bt *tenebrionis*(active against the Colorado potato beetle, *Leptinotarsa decemlineata*) (**Carlton and Gawron-Burke, 1993**).

Strain EG2348 (created by Ecogen) synthesizes a combination of Cry proteins particularly active against specific lepidopteran pests that infest soybean crops, and this is the active ingredient of Condor. It is not always possible to introduce plasmids by conjugation, particularly if this results in the coexistence of two plasmids with similar origins of replication in the same cell, as they would be incompatible.

In such cases, the relevant genes can be inserted onto a different plasmid in *Escherichia coli* using standard cloning methods and that plasmid can be introduced into an existing Bt strain by artificial transformation (**Arantes and Lereclus, 1991**).

Today, the Bt biopesticide market is dominated by Abbott Laboratories (Chicago, IL) (since the acquisition in 1995 of Novo-Nordisk's biopesticide business) and Novartis (created through the merger in 1996 of Ciba and Sandoz), together accounting for >70% of global production. The other 30% is divided among approximately 30 companies with over 100 different Bt product formulations, most containing a single Bt toxin but some combining up to five.

The bacteria produce a toxin in the insect's stomach that prevents food from being absorbed into its body. The insect stops feeding immediately. After several days, it dies from starvation. *BT* strains that affect caterpillars (*Bacillus thuringiensis kurstaki*) do not harm birds or beneficial insects that feed on caterpillars. *Bacillus thuringiensis israelensis* is used for certain fly larvae.

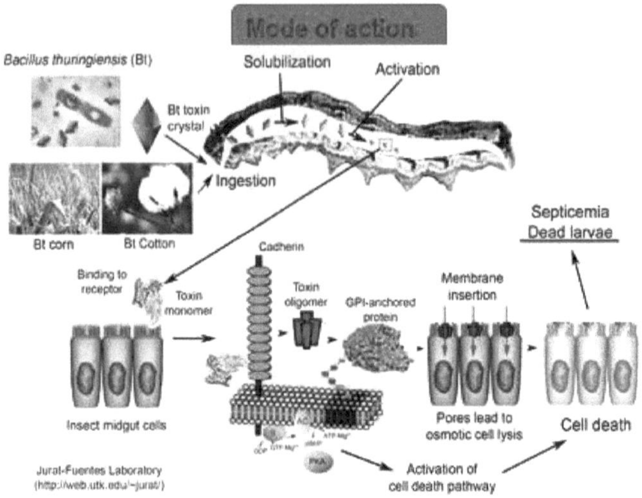

Mode of action
Bacillus thuringiensis (Bt)
Solubilization
Activation
Bt toxin crystal
Ingestion
Bt corn
Bt Cotton
Septicemia
Dead larvae
Cadherin
Binding to receptor
Toxin monomer
Toxin oligomer
GPI-anchored protein
Membrane insertion
Insect midgut cells
Pores lead to osmotic cell lysis
Cell death
Jurat-Fuentes Laboratory
(http://web.utk.edu/~jurat/)
Activation of cell death pathway

Bacillus thuringiensis (*Bt*) in transgenic cotton

- ❖ The *B. thuringiensis* strains produce three types of insecticidal toxins:
  - crystal (Cry)toxins,
  - cytolytic (Cyt) toxins and
  - vegetatively expressed insecticidal proteins (vip).
- ✓ Crystalline proteins are classified according to structure & have a specific nomenclature (e.g., Cry1Ac, Lepidoptera only).
- ✓ Total of 229 cry toxins (Cry1Aa to Cry72Aa), 11 cyt toxins (cyt1Aa to cyt3Aa) and 102 vip toxins (vip1Aa1 to vip4Aa1).
- ✓ A total number of 342 *Bt* toxin genes are available for research to develop insect resistant GM crops.
- ✓ These toxins are highly specific to certain insect species.

Why do we need *Bt* Cotton?

- ❖ Cotton is a long duration crop and is attacked by large number of insect pests throughout its growth and development.
- ✓ The three bollworms,

  Helicoverpa armigera,

  Pectinophora gossypiella &

  Earias insulana
- ❖ Insecticide quantity applied on cotton was the highest, relative to other cultivated crops.
- ❖ Cotton bolls are highly vulnerable to hidden insects .
- ✓ The pink and spotted bollworms are hidden feeders and generally do not come into direct contact with insecticide sprays.

How *Bt* Works?

- ✓ **Ingestion.**
- ✓ **Solublization**
- ✓ **Proteolytic activation.**
- ✓ **Binding to target site.**
- ✓ **Formation to toxic effects.**

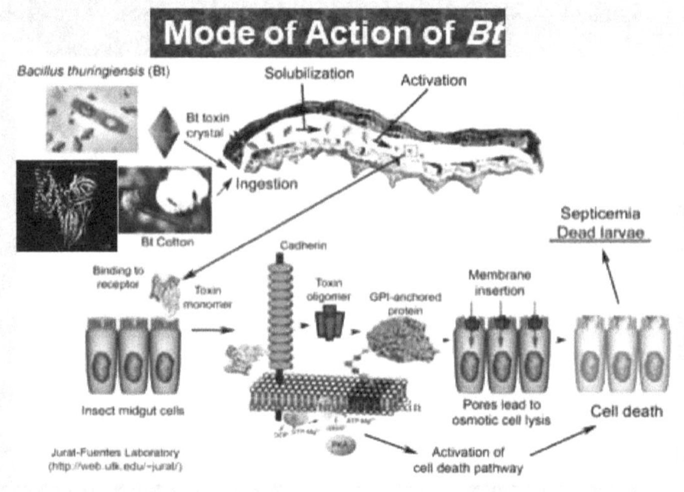

Four components of Transgenic Cotton

- **Laboratory phase.**
- **Breeding phase.**
- **Bio-Safety & Field studies.**
- **Regulatory Process for transgenic crops.**

Technology of Production

❑ Main steps for developing transgenic crops:

- ✓ Identification for effective gene(s).
- ✓ Gene transfer.
- ✓ Regeneration from protoplast/ callus/ tissue
- ✓ Gene expression to desired level.
- ✓ Back cross to produce varieties.
- ✓ Field test.
- ✓ Approval for commercialization.

Methods of gene transfer

- **Agrobacterium mediated gene transfer.**
- **Bombardment of cell with particle coated with gene of interest.**

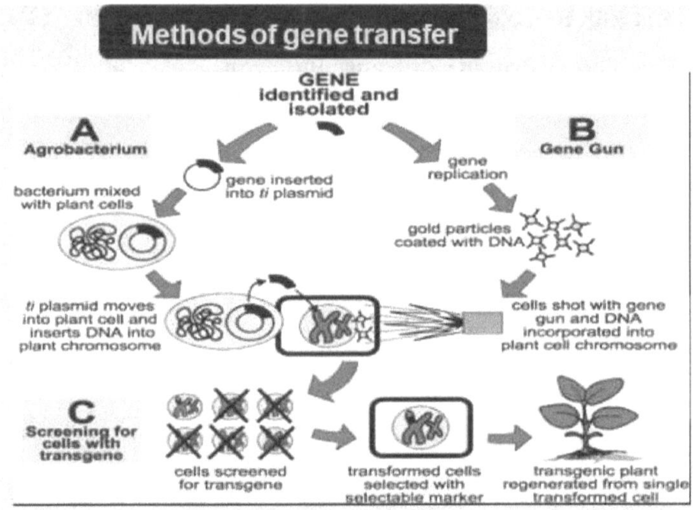

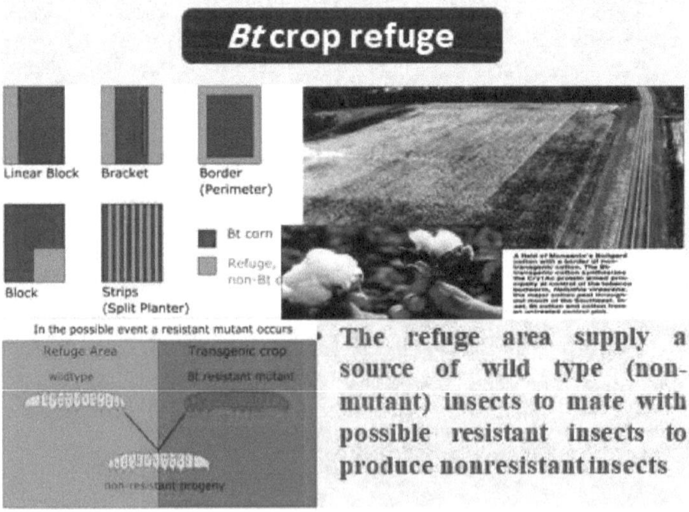

Development of Additional Transgenes (Bt's)

Bollgard I

1 *Bt* gene (Cry1Ac)

Bollgard II

2 *Bt* gene (Cry1Ac) + Cry2Ab

Bollgard III

3 *Bt* gene Cry1Ac + Cry2Ab + Vip

Advantages of *Bt* Cotton

Yield superiority

More profit

Lesser need of pesticide

Better quality

Suitability for early sowing

Disadvantages of *Bt* Cotton

Higher cost of seeds

Higher fertilizer and irrigation cost

Advantages and disadvantages of topical Bt pesticides

Topical Bt sprays are advantageous in terms of their safety, specificity and potency compared to chemical sprays, and are also biodegradable, which provides for a large and competitive market.

However, this last 'advantage' needs to be cited with caution, because Bt is only effective when present on the plant organs on which insects feed. Usually Bt is applied when early instar larvae are present, because older larvae are more tolerant. Bt sprays persist for only a few days on the leaf surface because UV light, weather, the chemical environment of the leaf surface and the presence of proteinases contribute to the degradation of Cry proteins.

Many spores are washed off the leaf surface into the soil. There is no evidence to suggest Bt is dangerous to humans and other mammals, and indeed the studies performed thus far suggest Bt is one of the safest microbial products known.

Given its extensive use, it is remarkable that only a single injury sustained from the use of Bt products in the field has been reported, the identification of Bt spores in the corneal ulcer of a farmer whose face was splashed with Dipel (**Burges, 2001**).

Laboratory mice can survive subcutaneous injections of 10^6 spores, or 10^7 spores administered intranasally (**Siegel, 2001**), and although limited mortality is evident when 10^8 spores are delivered intranasally, this is equivalent to an exposure level of 10^{12} in humans, or one billion times higher than the maximum ever encountered in the field during spraying.

Even so, concerns about the environmental persistence of Bt spores in soil and water have encouraged research into different formulations, including pure protein crystals that are applied in the same way as the spores. As these are even more vulnerable to degradation than the spores, Cry proteins have been encapsulated in the bacterium *Pseudomonas fluorescens* (e.g. in the Mycogen products MVP which targets lepidopteran pests and M-Trak which targets coleopteran pests).

This encapsulation strategy protects the Cry protein from UV light and chemical degradation, and allows large amounts of each Cry protein to be produced using high-yielding expression constructs, but the bacteria do not persist in soil or water for as long as Bt spores.

Crop Genetics International developed a similar strategy using *Clavibacter xyli* var. *cynodontis*, an endophytic bacterium that can penetrate the vascular system of plants, and this confers resistance to the European corn borer (**Lampel *et al.*, 1994**).

Bacillus thuringiensis sub species *israelensis* in mosquito control

Bt var *israeliensis* was isolated in Israel in 1976, and was shown to be effective against dipteran larvae including blackflies and mosquitoes (**Goldberg and Margalit, 1977**).

Therefore, whereas most Bt strains have been developed as topical pesticides for use in agriculture, Bt subsp. *israelensis* has been applied to water courses and stagnant pools to prevent the spread of malaria (**Margalith and Ben-Dov, 2000**; **Fillinger *et al.*, 2003**).

The toxins are encoded by a megaplasmid called pBtoxis which contains the genes for four Cry proteins (Cry4Aa, Cry4Ba, Cry10Aa, Cry11Aa) and two Cyt proteins (Cyt1Aa and Cyt2Ba) (**Berry *et al.*, 2002**).

The Cyt proteins interact directly with the lipid membrane in the larval midgut and act as receptors for the Cry toxins, so that the two act in synergy.

Product development has shown that formulations containing smaller particles are the most effective because they remain suspended for longer and are easily ingested by the filter-feeding larvae. Heavier particles sink and are covered in mud, quickly becoming ineffective, so specialist products that facilitate slow release have the best performance, e.g. products with the bacteria suspended in ice chips, or formulations including *Bacillus sphaericus*.

The development of Bt transgenic crops

Early development

One of the major disadvantages of topical Bt pesticides is their short window of effectiveness, and the fact that inclement weather can render Bt sprays useless within a matter of hours. Topical Bt sprays must therefore be reapplied several times in a growing season to reach the entire larval pest population, increasing the amount of product that needs to be used and the fuel needed for spraying. Furthermore, topical sprays have little impact on so-called cryptic pests, i.e. sap sucking insects and larvae that feed near the roots.

A potential solution to this problem was developed in the mid-1980s when scientists introduced Bt *cry* genes into tobacco and tomato plants and expressed the proteins directly in plant tissues.

The Belgian company Plant Genetic Systems pioneered commercial interest in Bt transgenic technology, but no Cry protein could be detected in the first generation of experimental transgenic plants (**Fischhoff *et al.*, 1987**; **Vaeck *et al.*, 1987**; **Perlak *et al.*, 1991**).

Experiments to determine the cause of low expression levels concentrated on differences between prokaryotic and eukaryotic systems because recombinant Cry proteins were expressed at high levels in heterologous bacteria.

A large difference in average GC content was noted between Bt and plant DNA, plus differences in codon preference that could account for low-efficiency protein synthesis. The bacterial genes also contained frequent ATTTA sequences, which in plants act as signals to increase the rate of mRNA turnover. **Perlak *et al.* (1991)** modified the sequence of the *cry1Ab* and *cry1Ac* genes to increase the GC content, replace sequences with four or more consecutive adenine or thymine residues and shift codon preference towards that favoured by plants, increasing protein levels by up to 100-fold and achieving total yields equivalent to 0.02% total soluble protein (TSP).

This was still insufficient for adequate pest control, but expression levels could be increased still further using stronger promoters, more efficient polyadenylation and termination signals, and by including a heterologous intron in the expression construct. The development of synthetic *cry* genes optimized for expression in plants meant that Cry proteins were soon expressed at levels of 0.2–1% TSP (**Koziel *et al.*, 1993**) and this increased to over 5% when *cry* genes were introduced into the chloroplast genome (**McBride *et al.*, 1995**).

Field trials and early commercial crops

After successful results in laboratory tests, the first field trials with Bt transgenic tobacco were conducted in the United States and France in 1986.

The plants expressed a truncated gene encoding the N-terminal (toxic) portion of Cry1A from Bt var *kurstaki* HD-73 under the control of the constitutive *Cauliflower mosaic virus* 35S promoter and protected the plants from leaf damage caused by *Helicoverpa zea*, a pest known variously as cotton bollworm, corn earworm or tomato fruitworm, depending on the crop it infests (**Hoffmann *et al.*, 1992**).

Whereas it was never likely that the Bt tobacco variety would be developed for commercial exploitation, transgenic potato plants expressing Cry3A from Bt var. *tenebrionis*were shown to protect the crop against Colorado potato beetle in the field much more efficiently than Cry3A topical sprays and were earmarked for commercial development (**Perlak *et al.*, 1993**). Trials with cotton, maize and rice soon followed (**Table 1**).

Table 1. Early field trials with Bt transgenic plants (data from Hilder and Boulter, 1999)

Crop	**Bt toxin**	**Target pest**
Tomato	Cry1A	Pinworm
Tobacco	Cry1A	*Helicoverpa zea*
Potato	Cry1A	Tuber moth
	Cry3A	Colorado potato beetle
Cotton	Cry1A	Pink bollworm
Maize	Cry1A	European corn borer
Rice	Cry1A	Stemborers

In 1995, the US Environmental Protection Agency (EPA) approved the first registration of Bt potato, corn and cotton crops. The first to reach the market was Monsanto's NewLeaf potato variety expressing Cry3A, swiftly followed by two transgenic corn hybrids expressing Cry1Ab to protect against the European corn borer, i.e. KnockOut by Syngenta (Basel, Switzerland) and NatureGard by Mycogen (both containing event 176).

Monsanto also released the cotton varieties Bollgard and Ingard (events 531, 757 and 1076) expressing a modified Cry1Ac toxin. Two additional Bt corn varieties expressing Cry1Ab were released shortly thereafter, namely Agrisure CB by Northrup King (event Bt11) and the widely discussed YieldGard variety (event MON 810) by Monsanto.

The early landscape of the biotech crop industry had thus been established (**Box 1**). NewLeaf potato and its successors (NewLeaf Y and NewLeaf Plus) were withdrawn from the market in 2002 (**Box 2**), and corn varieties containing event 176 were later withdrawn and replaced with more profitable products.

Growth and diversification of Bt crops

By 1998, the uptake of Bt crops had increased significantly as data became available showing the positive impact of Bt transgenic technology on agriculture and the environment (see below). In 1998, the EPA approved an insect-resistant tomato line (event 5345) expressing Cry1Ac, and in 2001 the Herculex corn variety jointly developed by Pioneer Hi-Bred (Johnston, IA) and Dow AgroSciences (Indianapolis, IN) (event TC 1507) expressing Cry1F and protecting plants against black cutworm (*Agrotis ipsilon*), fall armyworm (*Spodoptera frugiperda*) and the European corn borer. A new landmark was achieved in 2002, with the approval of Monsanto's Bollgard II cotton (event 15985), which expressed two Bt toxins, Cry1Ac and Cry2Ab, and later YieldGard Rootworm (event MON 863), expressing a synthetic variant of the *cry3Bb1* gene from Bt subsp. *kumamotoensis*, providing resistance against the western corn rootworm (*Diabrotica virgifera virgifera*).

The first stacked variety developed by crossing two previously released Bt varieties was also released in 2003. This was Monsanto's YieldGard Plus (event MON 810 + MON 863), which expressed Cry1Ab1 and Cry3Bb1. During this period, there was significant turbulence in the market as both large and small industry players manoeuvred to acquire strategic patents and technologies (**Box 3**). Tables S1–3 show, respectively, the current status of Bt patents, the companies currently engaged in the commercial development of Bt and the commercial status of different Bt crops.

Box 1 Portrait of an industry

The early pioneers of commercial Bt technology were agrobiotechnology companies such as Monsanto, DuPont, Calgene and Agracetus, which had collaborated with academic research groups to carry out field trials and develop potential commercial products in the mid-to-late 1980s.

As the 1990s approached, these companies entered into agreements with major seed distributors such as Delta and Pine Land (DPL) to develop and test their enhanced transgenic varieties. The predominant role of Monsanto in the early commercial landscape reflected their exclusive agreement with DPL to market transgenic cotton seeds internationally. Later, Monsanto forged agreements with several local companies in developing countries particularly China and India (Huang *et al.*, 2002a,b).

Early in the 1990s, Monsanto purchased Dekalb Genetics, Asgrow and Holden's acquiring elite germplasm through which to commercialize its biotechnology products.

Evolution of the commercial landscape for Bt crops. The five major companies currently selling Bt seeds have arisen through a series of mergers, acquisitions and spin offs/demergers as larger companies segregate their agribusiness interests.

Monsanto Co., in its current incarnation, was an agribusiness spin-off from Pharmacia in 2002 following the merger of the original Monsanto Co. (established in 1901) with Pharmacia and Upjohn in 2000. Pharmacia created the new Monsanto as an agribusiness subsidiary in late 2000, and then established it as an independent company in 2002.

Bayer CropScience is an agribusiness subsidiary of Bayer AG, formed following the acquisition of Aventis CropScience in 2000. Syngenta formed from the merger of Novartis and AstraZeneca in 2000, both of which were agribusiness spin-offs generated in previous mergers. Dow AgroSciences is a wholly owned subsidiary of Dow Chemical Co., formed when Dow Chemical Co. purchased Eli Lilly's stake in Dow Elanco (an agribusiness spin-off from Dow Chemical Co. and Ely Lily & Co. formed in 1989).

Finally, Pioneer Hi-Bred International is now an agribusiness subsidiary of DuPont, which acquired 20% of the company in 1997 and the remaining 80% in 1999.

The Monsanto business model of assimilating companies with aligned strategic objectives was soon replicated by the other major players.

Although the 1990s was characterized by aggressive positioning, mergers and acquisitions, this has now given way to a more collaborative industry. The recent approval of SmartStax corn, codeveloped by Monsanto and Dow AgroSciences is one example, and others include the Greenleaf Genetics collaboration between Syngenta and Pioneer Hi-Bred International, and Monsanto licensing RoundupReady events to Syngenta so that the trait can be incorporated into their hybrids (**Marra *et al.*, 2010**).

International expansion

The United States has enthusiastically embraced the development of Bt agriculture and has by far the largest amount of land in total used for Bt or Bt stacked crops. However, Bt crops are grown in 25 other countries and the number of countries adopting Bt crops and the amount of land set aside for their cultivation has shown a continued upward trend for 15 years.

Bt agriculture is expanding on every continent except Europe, which persists with its absurdly byzantine approach to all genetically engineered crops. Remarkably, the small African nation of Burkino Faso grows more Bt crops than the whole of Europe. The total global area devoted to Bt crops in 2009 was >50 million hectares (36% of all biotech crops), made up of 21.7 million hectares of Bt-only crops and 28.7 million hectares of crops with Bt stacked with herbicide tolerance (**James, 2010**).

Sharing out the Bt pie. More than 50 million hectares of Bt crops were grown commercially in 2008, the vast majority (>33 million hectares) in the USA. India, China, Argentina, Brazil, South Africa, Canada, Philippines, Australia and Uruguay were (in descending order by land area) the other countries to grow >100 000 hectares. The minor growers (in descending order by land area) were Spain, Mexico, Colombia, Honduras, Burkino Faso, Czech Republic, Romania, Portugal, Germany, Poland, Slovakia and Egypt. The data include Bt-only crops and Bt stacked with other traits. Source: **Brookes and Barfoot (2010)**. Although Argentina and Brazil currently hold second and third place in the global rankings for Bt agriculture, China and India have seen the most rapid adoption.

This is because both are major growers of cotton, and China in particular is a major grower of rice. Field trials of Bt rice were first conducted in China in 1998. A series of transgenic Bt rice lines transformed with modified *cry1A*, *cry1Ab* and *cry1Ac* genes were assessed in large-scale trials in 2007 (**Huang *et al.*, 2007**) and were approved for commercial release in November 2009, although large-scale cultivation is still pending.

NatureMark was a subsidiary of Monsanto created by the company in 1996 to market its new transgenic potato lines, beginning with NewLeaf (containing the gene for Cry3Aa) and subsequently NewLeaf Y (containing the Bt gene and an additional gene conferring resistance to *Potato virus Y*) and NewLeaf Plus (containing the Bt gene and an additional gene conferring resistance to *Potato leaf roll virus*). Despite rapid take up and hugely favourable responses from growers, all the potato lines and the NatureMark brand itself were abandoned in 2002 after a highly successful potato growing season in 2001. Why did this happen?

NewLeaf Plus field (left) is protected from *Potato leaf roll virus* (PLRV) without sprays. No infection is evident. The conventional potato field on the right is 100% PLRV infected, despite sprays. Figure used with permission from Peter Thomas.

The NatureMark story shows how misguided activism can have disastrous effects. Activists began a very public campaign of misinformation against Monsanto's potato products in 1999 and soon had the support of several major convenience food providers including McDonalds, which banned genetically engineered crops from its food to avoid negative publicity.

This in turn put pressure on potato processers and growers, who had little choice but to cancel contracts for transgenic potatoes they would not be able to sell. The transgenic varieties had all received appropriate clearance from the regulators and were safe for human consumption, so it is clear that the entire chain of events was based on hysteria and not one shred of scientific evidence.

Notwithstanding the above, Monsanto was forced to abandon NatureMark and all potato-related research and development as there would be no market for its products. The activists were successful, but it was a Pyrrhic victory.

In their haste to save the environment by getting rid of 'unnatural' vegetables, they ensured that tons of chemical pesticides would be required for future potato crops and tons of fuel would be required to spray them; they ensured that farmers in North America would mourn the loss of a superior product, lose profit and make the agricultural sector as a whole less profitable; and they ensured that many additional beneficial crops would never see the light of day, or would be developed instead in other countries.

Box 3 War and peace

Many aspects of Bt technology fall under the scope of intellectual property and can be protected with patents, although many of these aspects (processing and formulation in the case of Bt sprays, gene transfer and expression in the case of Bt crops) are not specific to Bt and can have wider ramifications.

Many companies and academic departments working on Bt technology have sought patents, and in 1996 when the first Bt crops were commercialized these were divided more or less equally between the 'old guard' companies that had developed Bt topical products and the 'new wave' of companies expressing Bt genes in plants.

Nearly 60% of the approximately 400 Bt patents were owned by Mycogen, Novartis, Abbot, Toa, AgrEvo, Ecogen, Monsanto and Zeneca. It is noteworthy that the five major companies involved in Bt crop development today have, since 1996, engaged in many cooperative research and development agreements and this has continued to the present day. However, they have also litigated extensively to prevent patent infringements.

Ecological aspects of Bt crops

Potential for the evolution of resistant insect populations

All insecticides create selection pressure on target populations, and the mode of action of Bt toxins (binding to a specific receptor on midgut epithelial cells) presents a clear opportunity for pests to evolve resistance. The first evidence of this process was observed in 1985, when resistant mealmoths (*Plodia interpunctella*) were found in grain stores that had been sprayed with Bt spores. The selection pressure was recreated in the laboratory, showing the evolution of resistant strains after 15 generations of sublethal selection (**McGaughey, 1985**).

Resistance was also observed in wild populations of the diamondback moth (*Plutella xylostella*) feeding on watercress in Hawaii that had been sprayed with Bt up to 400 times (**Liu and Tabashnik, 1997**).

Laboratory experiments were able to produce Bt-resistant varieties of several additional species that had not evolved resistance in the wild, suggesting that the intensive use of single Cry proteins was likely to result in the evolution of resistant strains (**Tabashnik *et al.*, 2005**).

The Bt crop industry is aware of the danger of resistant pests, and many seed vendors insist on customer agreements that mandate the use of preventative measures, particularly the refugia strategy in which a proportion of any field containing Bt crops must be planted with the nontransgenic variety to encourage the breeding of nonresistant pests.

The widespread use of this strategy is probably responsible for the remarkable lack of resistant populations even in areas devoted to high-intensity Bt agriculture for 15 years.

Tabashnik and coworkers have studied pest populations on Bt sites in the United States, Australia, China and Europe and have found that among six major insect pests, field resistance occurred in only one species (*H. zea*) and only at a limited number of sites in Arkansas and in Mississippi, not in, for example, North Carolina, where the refuge areas are typically larger (**Tabashnik *et al.*, 2008**).

The prolonged efficacy of the first generation of Bt crops for more than a decade against nearly all targeted pest populations has exceeded the expectations of many entomologists working on population genetics (**Bourguet, 2004**). Although integrated pest management strategies have been carefully implemented by growers, the absence of resistant populations in the wild suggests that resistance may attract a fitness penalty in the absence of the toxin (**Sanchis and Bourguet, 2008**).

Although current refugia strategies have worked better than anticipated, a number of alternative or complementary approaches have been proposed to address the possibility of resistance evolving in target pests.

High-dose and low-dose approaches aim in one case to overwhelm insect populations with immense toxin doses so that there is no chance to evolve resistance, and in the other case merely to make them more susceptible to predators, but both make many assumptions and would be difficult to implement in the field (**Gould, 1994**).

Similarly, targeted expression (e.g. wound-inducible expression) or temporally restricted expression would provide the same advantages of a refugia but would be more difficult to implement. A more robust approach is resistance pyramiding, i.e. the stacking of multiple genes in the same plant that target the same pest via different mechanisms. In theory, resistance against one Cry protein could arise through a single point mutation in the gene encoding its receptor.

Experiments have shown that mutations affecting the secondary, high-affinity Cry toxin receptors do not induce resistance, but multiple resistance alleles can be identified in the cadherin genes encoding the primary receptors (**Soberón *et al.*, 2009**; **Zhao *et al.*, 2010**).

However, the chances of two mutations arising simultaneously in the receptors for two independently acting toxins would be much lower. This is probably why mosquito strains resistant to Bt var *israelensis* have not evolved despite many years of deployment—the toxin crystals in this case comprise five different toxins, one of which is Cyt1A which, as discussed above, has a different action mechanism to the three-domain Cry toxins. Pyramiding resistance crops are therefore likely to require smaller refuges (**Shelton *et al.*, 2002**).

The pyramiding strategy is supported by laboratory tests with cotton bollworm and the recent analysis of pink bollworm (*Pectinophora gossypiella*) populations in Bt cotton fields in India. The laboratory studies showed that cotton bollworm can evolve resistance to MVP, a commercial Bt formulation based on Cry1Ac, but not to DiPel or XenTari, which contain multiple Cry proteins (**Akhurst *et al.*, 2003**).

The resistant strain was also resistant to Cry1Ab (which binds to the same receptor as Cry1Ac in cotton bollworm) but not to Cry2Aa or Cry2Ab which bind different receptors (**Liao *et al.*, 2002**).

Field monitoring of pink bollworm in 2009 showed that the pest evolved resistance to Bollgard I cotton (Cry1Ac) in four areas of Gujarat: Amreli, Bhavnagar, Junagarh and Rajkot. However, no resistance was observed in fields growing the Bollgard II variety, which contains Cry1Ac and Cry2Ab (**Monsanto, 2010**).

Attempts have also been made to enhance toxin activity by coexpression or protein fusion. One example is the expression of Cry toxins along with fragments of their receptors, which can potentiate their activity by allowing them to assemble into pore-forming complexes immediately (**Chen *et al.*, 2007**).

The fusion of two or more toxins has been used to generate artificial hybrids with a host range differing from that of the parent toxins. For example,**Naimov *et al.* (2003)** created a fusion toxin comprising a truncated version of Cry1Ba and domain II from Cry1Ia. Desiree potato plants expressing this recombinant toxin were the first Bt plants resistant to both coleopteran and lepidopteran pests [Colorado potato beetle, potato tuber moth (*Phthorimaea operculella*) and European corn borer] and the hybrid protein did not compete for binding sites with either of the parent toxins, indicating it bound to a distinct receptor. More recently, **Walters *et al.* (2010)** created a hybrid Cry1Ab/Cry3A toxin (eCry3.1Ab) which was toxic to the western corn rootworm, a pest unaffected by either of the parent toxins. In a related approach, **Mehlo *et al.* (2005)** fused the Cry1Ac toxin to the galactose-binding domain of the nontoxic ricin B-chain, again expanding its repertoire of potential receptors and therefore broadening its host range.

Transgenic rice and maize plants expressing the fusion protein were significantly more toxic in insect bioassays than those containing the Bt gene alone. They were also resistant to a wider range of insects, including important pests that are not normally susceptible to Bt toxins.

Environmental impact

Although there is much debate both politically and publically concerning the environmental impact of genetically engineered crops, it is clear that Bt crops have provided immense environmental benefits. The deployment of Bt crops has reduced the use of pesticides, also saving on fossil fuels required for spraying, reducing CO_2 emissions by limiting the need for ploughing, and conserving soil and moisture by encouraging no-tilling agriculture.

The cumulative reduction in pesticide use for the period 1996–2008 was approximately 356 000 tonnes (8.4%), which is equivalent to a 16.1% reduction in the associated net environmental impact as measured by the environmental impact quotient (EIQ). The corresponding data for 2008 alone revealed a reduction of 34 600 tonnes of pesticides (9.6%) and a reduction of 18.2% in EIQ (**Brookes and Barfoot, 2010**).

In countries such as India, China, Argentina and Brazil, which are the most enthusiastic adopters of Bt agriculture after the United States, the greatest impact of Bt has been the reduction in the number of pesticide sprays (from 16 down to 2–3 per growing season) and a concomitant reduction in the number of poisonings caused by chemical exposure. These factors, together with average yield increases of up to 10%, have raised net income by as much as 40% (**Subramanian and Qaim, 2010**).

Although the reduction in pesticide use has been beneficial to the environment and the economy, concern has been expressed that the use of Bt transgenic plants could affect beneficial insects, upset the balance of natural ecosystems and encourage the breeding of secondary pests. It is unclear why those raising such concerns are not equally concerned about the impact of topical Bt sprays and chemical pesticides on these populations, because sprays are the only current alternative to transgenic crops that will guarantee adequate food production.

Beneficial insects

The potential impact of Bt crops on beneficial insects was brought into focus by the now discredited Monarch butterfly study, which suggested Monarch larvae feeding on leaves covered in pollen shed from Bt maize plants (event Bt176) did not grow as rapidly as those feeding on uncontaminated leaves. This report was seized on by opponents of genetic engineering technology and is still routinely cited as an argument against the deployment of Bt crops despite follow-up studies finding no evidence for a statistically significant effect. What other evidence is there regarding the impact of Bt crops on nontarget insects?

Field studies on the NewLeaf potato (Cry3Aa) showed that the toxin specifically affects the Colorado potato beetle and has no deleterious effect on other insects in the potato field, including the beetle's natural predators.

In contrast, chemical sprays killed both the beetle and its predators, leading to an explosion in the population of vectors carrying viral pathogens, thus increasing the risk of potato virus diseases (**Reed *et al.*, 2001**).

Any impact on natural predators that normally keep pest populations in check could have knock-on effects throughout the food web, so careful studies of these effects are required (**Dutton*et al.*, 2002**).

One such study looked at nontarget arthropod predators in Bt maize fields (specifically events MON 810 expressing Cry1Ab, MON 88017 expressing Cry3Bb1 and a stacked variety MON 89034 × MON 88017, expressing Cry1A105, Cry2Ab2 and Cry3Bb1).

The study showed that the predator and alternative prey populations naturally adjusted to reflect the absence of the targeted pest (**Faria *et al.*, 2007**).

Bt maize increased the population of corn aphid (*Rhopalosiphum maidis*) which resulted in more honey dew synthesis, which increased the number and longevity of the lepidopteran larval parasitoid *Cotesia marginiventris*. Bt cotton appears to have no effect on the cotton aphid (*Aphis gossypii*) population, and the Bt toxin was not detected in the honey dew, which is an energy source for many arthropod species including predators and parasitoids. Bt cotton therefore has no negative impact on beneficial insects in the cotton ecosystem (**Lawo *et al.*, 2009**).

Secondary pests

A secondary pest is a pest species whose numbers are usually kept in check by the presence of a primary pest, such that no control measures are necessary. However, elimination of the major pest may elevate the secondary pest to primary status, perhaps even affecting surrounding crops that are not usually troubled by either the primary or secondary pest species.

Cotton bollworm is a primary pest of cotton, and it suppresses the population of mirid bugs, i.e. homopteran insects that feed on plant sap. Bt cotton represents approximately 95% of all cotton in Northern China and is lethal to the cotton bollworm at the larval stage, so a study was carried out to look at any impact on mirid bug populations (**Lu *et al.*, 2010**).

The study showed that mired bug populations have not increased in nontransgenic cotton because the species is controlled by broad-spectrum pesticides that are also used to kill bollworm larvae. In Bt cotton, the mirid bug population has increased every year from 1997 to 2008 and has gained the status of a primary pest, a phenomenon that is now impacting on unrelated crops such as dates, grapevine, apple, peach and pear. Although this is an undesirable outcome, it is somewhat balanced by the increased insect biodiversity observed in Bt cotton in China.

Field studies revealed 31 insect species in Bt plots (23 beneficial) compared to 14 species in non-Bt plots, and only five of which were beneficial (**Pray *et al.*, 2002**).

Environmental diversity

In addition to insect populations, it is useful to study the impact of Bt on other parts of the ecosystem, particularly the soil as this is where Bt spores end up when washed from the plant surface, and is the destination of Bt toxins exuded from plant roots, released from pollen grains and released from decaying or residual plant biomass ploughed into the soil. Earthworms (oligochaetes) are good indicators of general soil health and comparisons of earthworm numbers in plots containing nontransgenic maize and Bt maize expressing*cry1Ab* (events Bt11 and MON 810) and *cry3Bb1* (event MON 863) over 4 years showed no differences in development or biomass (**Zeilinger *et al.*, 2010**).

More earthworms were found within the rows of maize plants than between them in all plots, perhaps because the soil is lighter and has more biological activity and therefore represents a better source of nutrients.

Looking to the future

The first generation of Bt crops has been extraordinarily successful, with a few examples of pest populations evolving resistance. These crops are already being supplanted with second-generation varieties with more resilient traits generated by stacking and pyramiding resistance genes.

Even so, this is not the time to be complacent and the search for more efficacious and potent strains must continue (**Christou *et al.*, 2006**; **Crickmore, 2006**). New strains of Bt are reported on a regular basis, especially now proteomics methods can be used to screen for novel toxins on a large scale. **Sun and Park (2010)**recently identified Cry60Ba from Bt serovar *malayensis* in a search for mosquitocidal toxins, and **Zhang *et al.* (2010)** recently isolated strain LLP29 from the phylloplane of *Magnolia denudate*, identifying a novel toxin (Cyt1Aa6) which is lethal to mosquito larvae. Homologous genes have also been identified in related bacteria, reflecting the fact that Cry proteins are members of a diverse superfamily.

Strain engineering efforts have continued to extend the Bt host range. For example, **Liu*et al.* (2010)** recently reported the construction of strain BIOT185 from the original strains HBF-1 and BTO 185 that express Cry8ca2 and Cry8Ea1, respectively.

The new strain is toxic towards scarab insects such as *Atractaspis corpulenta*. Similarly, **Wang *et al.* (2008)**constructed a new strain by introducing the *cry3Aa7* gene into the UV17 strain, which produces Cry1Aa, Cry1Ac, Cry1Ca and Cry2Ab. The new strain was toxic to both lepidopteran and coleopteran insects.

The toxicity of Cry proteins can be enhanced by amino acid substitutions, the introduction of cleavage sites in specific regions of the protein and the deletion of small fragments from the N-terminal region (**Pardo-López *et al.*, 2009**).

For example, replacing residue N372 in Cry1Ab domain II with alanine or glycine increases its toxicity to the gypsy moth (*Lymantria dispar*) eightfold by inducing a fourfold increase in binding affinity to its receptor (**Rajamohan *et al.*, 2006**).

More recently, **Muñóz-Garay *et al.* (2009)** produced an engineered Cry1AMod toxin lacking helix α-1, which did not need to bind the receptor cadherin and therefore killed even insects that were resistant to the parent toxin, Cry1Ab.

The commercial environment for stacked and pyramiding traits in Bt crops was given a boost in 2010 with the approval of SmartStax corn, codeveloped by Monsanto and Dow AgroSciences. Approvals were granted in the United States, Canada, Japan, Mexico, The Philippines, Taiwan and South Korea (**Marra *et al.*, 2010**). SmartStax was created by crossing several existing varieties to stack all the events (MON 89034 × TC 1507 × MON 88017 × DAS-59122-7) in one line. SmartStax corn contains eight transgenes, including six Bt genes offering a broad spectrum of pest resistance above and below ground, while still retaining the specificity of each toxin. The specific toxins are Cry1A.105, Cry2Ab, Cry3Bb1, Cry34Ab1, Cry35Ab1 and Cry1Fa2, providing resistance to a long list of coleopteran and lepidopteran pests, including the European corn borer, south-western corn borer (*Diatraea grandiosella*), southern cornstalk borer (*Diatraea crambidoides*), corn earworm, fall armyworm, stalk borer (*Papaipema nebris*), lesser corn stalk borer (*Elasmopalpus lignosellus*), sugarcane borer (*Diatraea saccharalis*), western bean cutworm (*Striacosta albicosta*), black cutworm, western corn rootworm, northern corn rootworm (*Diabrotica barberi*) and Mexican corn rootworm (*Diabrotica virgifera zeae*), combining this with tolerance to two different broad-spectrum herbicides (glyphosate and glufosinate). The refuge requirement for SmartStax corn is just 5%.

Nine common pests of rice, cotton and maize that are controlled by Bt crops. Top row from left (rice pests): yellow stem borer (*Scirpophaga incertulas*), rice leaffolder (*Lerodea eufala*), striped stem borer (*Chilo suppressalis*). Middle row from left (cotton pests): pink bollworm (*Pectinophora gossypiella*), tobacco budworm (*Heliothis virescens*), American bollworm (*Helicoverpa armigera*). Bottom row from left (maize pests): European corn borer (*Ostrinia nubilalis*), fall armyworm (Spodoptera frugiperda), northern corn rootworm (*Diabrotica barberi*). Image sources from top left: (i) IRRI, (ii) IRRI, (iii) Bayer CropScience, (iv) USDA, (v) Cotton Corporation of India Ltd., (vi) USDA, (vii) Frank Peairs, Colorado State University, Bugwood.org, (viii) Canadian Biodiversity Information Facility, (ix) Plant Management Network.

The first 100 years of Bt topical sprays and transgenic crops have been extraordinarily successful and advantageous, with a strong record in terms of safety, efficacy and environmental beneficence, and the ability to bring economic prosperity in both developing countries and the industrialized west. Recent reports show that work continues to identify and create novel Bt strains and toxins with more potent and specific effects, and to generate transgenic plant lines that suppress agricultural pests and reduce opportunities for the evolution of resistant strains while conferring no harm on beneficial insects or soil organisms.

This unique combination of benefits will continue to provide farmers and health professionals with the weapons they need to fight pests and insect-borne diseases, maintaining crop yields and improving the health of the world's growing population.

Properties

Susceptible insects must ingest Bt toxin crystals in order to be affected. In contrast to poisonous insecticides that target the nervous system, Bt acts by producing a protein that blocks the digestive system of the insect, effectively starving it. Bt is a fast-acting insecticide: an infected insect will stop feeding within hours of ingestion and will die, generally from starvation or a rupture of the digestive system, within days.

Whether applied in spray form or through genetic engineering, each Bt strain is effective against a narrow range of insects. The most commonly used strain of Bt (*kurstaki*, or Btk) targets only certain species of caterpillars. Since the late 1970s, Bt strains (e.g.,*israelensis*, or Bti) have been developed that control certain types of fly larvae, including those of mosquitoes, black flies, and fungus gnats. Other common strains include *san diego and tenebrionis*, which are effective against certain leaf beetles, such as theColorado potato beetle and elm leaf beetle.

Advantages

Unlike most insecticides, which target a broad spectrum of species, including both pests and beneficial insects, Bt is toxic to a narrow range of insects. Research suggests that Bt does not harm the natural enemies of insects, nor does it impair honeybees and other pollinators critical to agroecological systems. Bt integrates well with other natural controls and is used for integrated pest management by many organic farmers.

The use of insect-resistant Bt plants can potentially reduce use of chemical insecticide sprays, which are extremely toxic and expensive. Applications of conventional pesticides recommended for control of the European corn borer, for example, dropped by about one-third after Bt corn was introduced.

Although lethal to certain insect species, Bt toxin applied as an insecticide or consumed with GMO food crops is considered nontoxic to humans and other mammals because they lack the digestive enzymes needed to activate the Bt protein crystals. However, any introduction of new genetic material is potentially a source for allergens, and, for this reason, certain strains of Bt are not approved for human consumption.

Disadvantages

Bt, when applied in spray or liquid form, is susceptible to degradation by sunlight. Most formulations persist on foliage less than a week following application. Some of the newer strains developed for leaf beetle control become ineffective in about 24 hours. Additives, such as sticking or wetting agents, are often useful in a Bt application to improve performance, allowing it to cover foliage more thoroughly and to resist washing off.

Given that Bt crops kill a narrow spectrum of the insects that prey upon them, additional insecticides are often required to protect the plants from damage by other pests. Additionally, the potential for insects to develop a resistance to the toxin as a result of repeated exposure is of great concern in the widespread cultivation of Bt crops.

Such resistance would render useless one of the most environmentally benign insecticides in use today. Already, certain moth and cotton pest populations have acquired resistance. Risk management strategies include the planting of refuges, such as a plot of non-Bt corn near a field planted in Bt corn, in order to maintain a local population of insects that remain susceptible to Bt toxin.

There is some degree of uncertainty about the impact of Bt proteins on the environment. According to the EPA, more research is needed to study Bt protein accumulation in soils, the risks posed to nontarget organisms, and the likelihood of gene flow from Bt crops to wild relatives.

A 2003 study from the Ohio State University, in which wild sunflowers were experimentally cross-pollinated with genetically modified Bt sunflowers, suggests that modified genes in cultivated crops may drift into closely related populations and increase the hardiness of these plants, including potential weed species.

Entomopathogenic nematodes

Entomopathogenic nematodes are soil-inhabiting, lethal insect parasitoids that belong to the phylum Nematoda, commonly called roundworms.

The term *entomopathogenic* comes from the Greek word *entomon,* meaning insect, and pathogenic, which means causing disease.

Although many other parasitic nematodes cause diseases in plants, livestock, and humans, entomopathogenic nematodes, as their name implies, only infect insects.

- Entomopathogenic nematodes (EPNs) live inside the body of their host, and so they are designated *endoparasitic.*
- They infect many different types of soil insects, including the larval forms of butterflies, moths, beetles, and flies, as well as adult crickets and grasshoppers.
- EPNs have been found in all inhabited continents and a range of ecologically diverse habitats, from cultivated fields to deserts.
- The most commonly studied genera are those that are useful in the biological control of insect pests, the Steinernematidae and Heterorhabditidae (Gaugler 2006).

Biological Control (BC) is an essential part in integrated pest management (IPM) systems, since it is one of the rare alternatives to the use of chemical pesticides.

IPM is a complex approach to pest management incorporating human, environmental and economical aspects into a sustainable pest management program.

An IPM system combines different disciplines, resources and management strategies in a multilateral integrated system (Cuperus *et al.,* 2000).

Entomopathogenic nematode is the kind of beneficial nematodes .it is divided into *Steinernema and Heterorhabditis sp.* it is easily isolated from the soil and has effective role in management the insects. Additionally, Entomopathogenic nematode can multiply insides the body of the insects and after this step it releases again into the soil to complete its cycle to invade and manage another enemy.

Life cycle

- Since they are economically important, the life cycles of the genera Heterorhabditidae and Steinernematidae are particularly well known.
- Although not closely related, phylogenetically, both share similar life histories (Poinar 1993)

- The cycle begins with an infective juvenile, whose only function is to seek out and infect new hosts.
- After entering an insect, infective juveniles release an associated mutualistic bacterium.
- These bacteria of the genus *Xenorhabdus* or *Photorhabdus*, for steinerernematides and heterorhabditids, respectively—cause host mortality within 48 hours.
- The nematodes provide shelter to the bacteria, which, in return, kill the insect host and provide nutrients to the nematode.
- Together, the nematodes and bacteria feed on the liquefying host, and reproduce for several generations inside the cadaver.
- Steinernematid infective juveniles may become males or females, whereas heterorhabditids develop into self-fertilizing hermaphrodites with later generations producing two sexes.
- When food resources in the host become scarce, the adults produce new infective juveniles adapted to withstand the outside environment.
- After about a week, hundreds of thousands of infective juveniles emerge and leave in search of new hosts, carrying with them an inoculation of mutualistic bacteria, received from the internal host environment (Boemare 2002, Gaugler 2006).

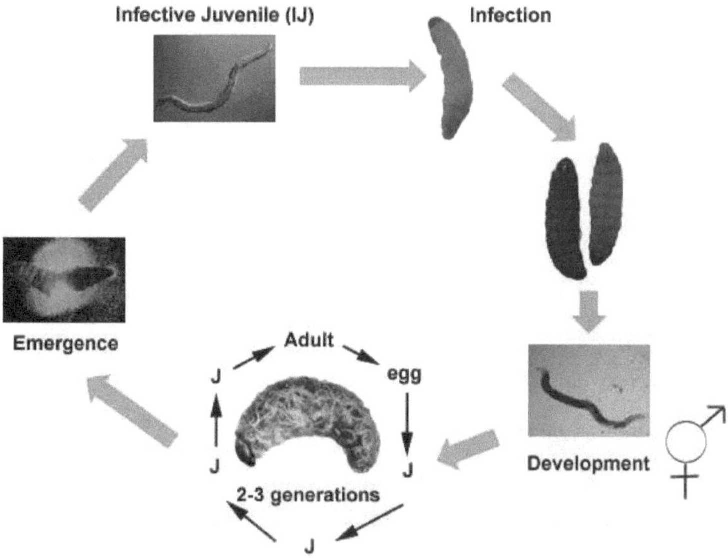
Infective Juvenile (IJ)
Infection
Emergence
Adult
egg
J
J
J
J
2-3 generations
Development

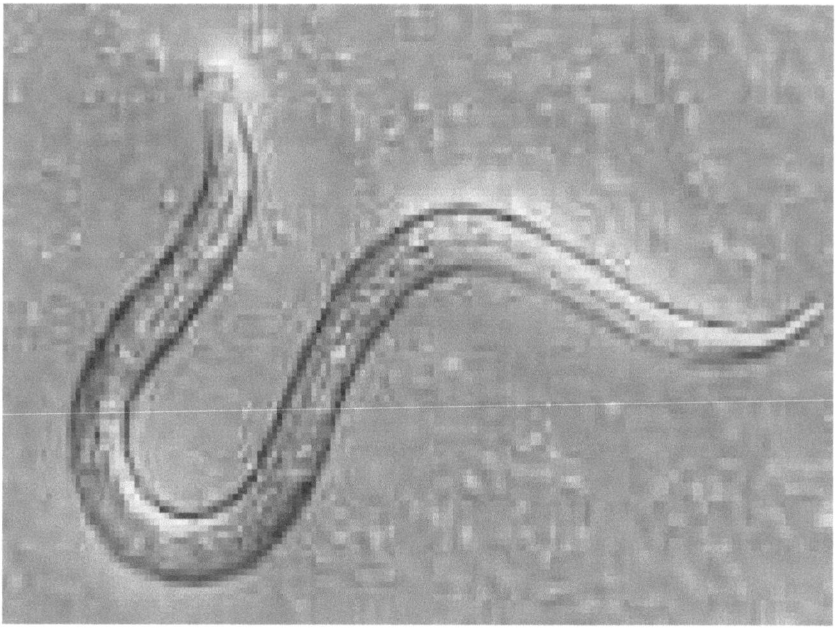

Tens of thousands of infective juveniles of the entomopathogenic nematode Steinernema carpocapsae spilling out of a Galleria wax moth larva in search of new hosts (photo provided by Randy Gaugler).

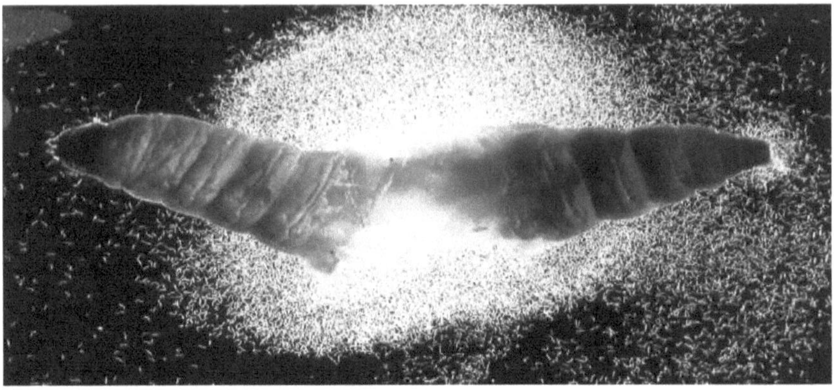

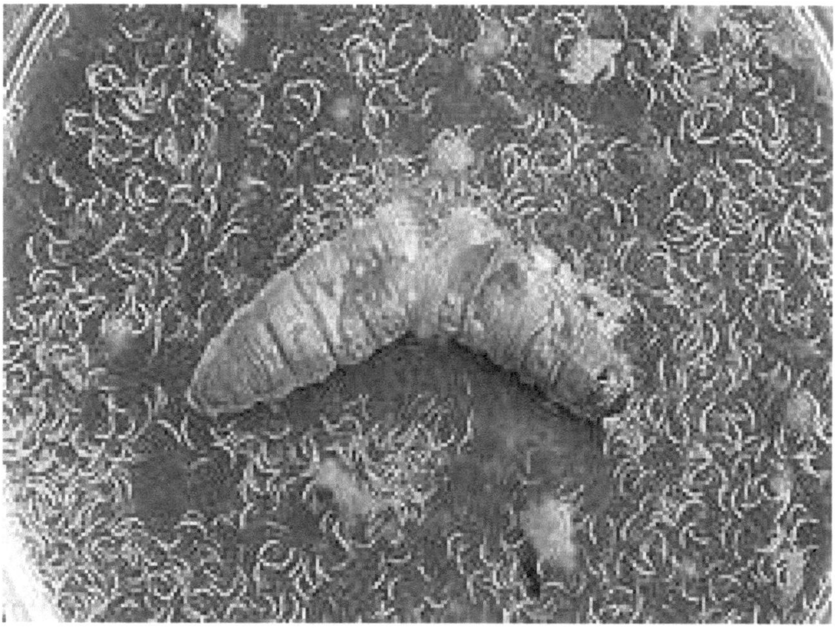

An infective juvenile entomopathogenic nematode

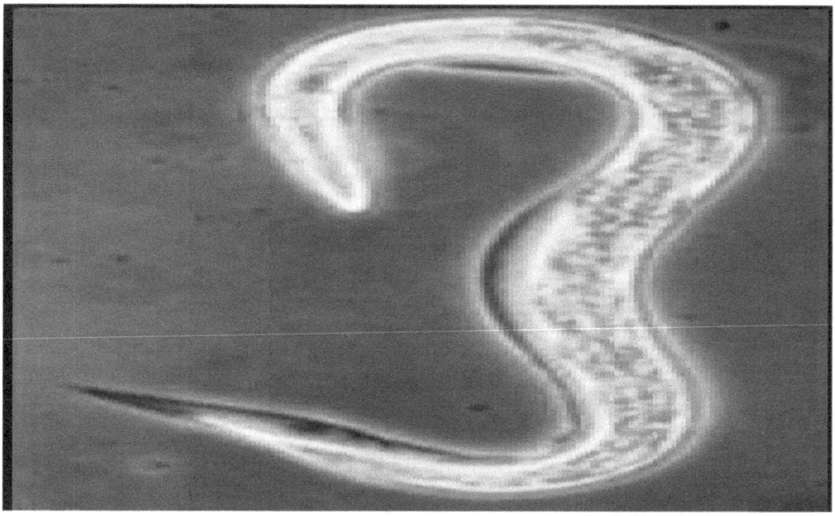

Life cycle of *Photorhabdus* and *Xenorhabdus*

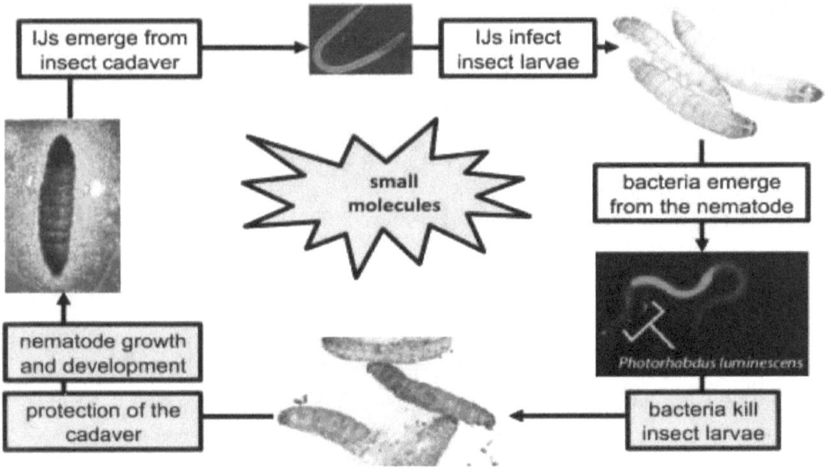

EPNs are associated with symbiotic bacteria in family enterobacteriaceae (genera *Xenorhabdus and Photorhabdus*).

The bacteria from both genera are gram-negative and belong to family Enterobacteriaceae.

The bacteria are stored in the anterior part of the IJ intestine in case of *Steinernema* and along the alimentary canal in *Heterorhabditis.*

- Their number can vary between 0 to 2000 cells/IJ. The bacteria are highly virulent and cause rapid death of its host.
- Once the bacteria are released from IJ into the haemocoel of an insect, they cause septicaemia, and in that way are capable of killing a pest within 48 h.

Secondary metabolites effect of symbiotic bacteria associated with EPNs on *G. mellonella*

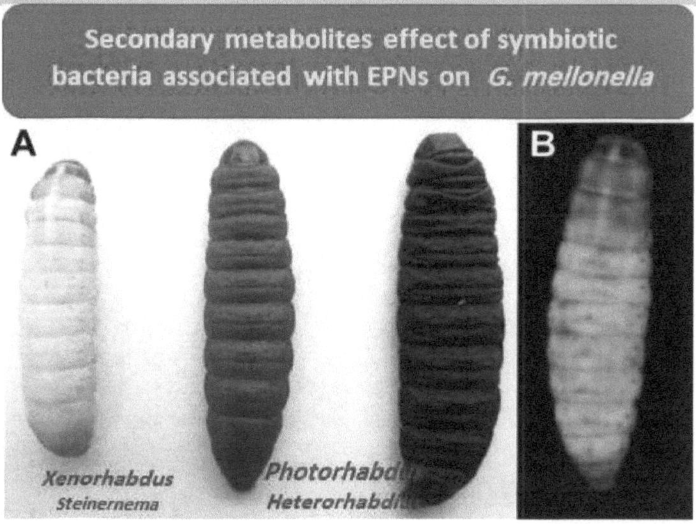

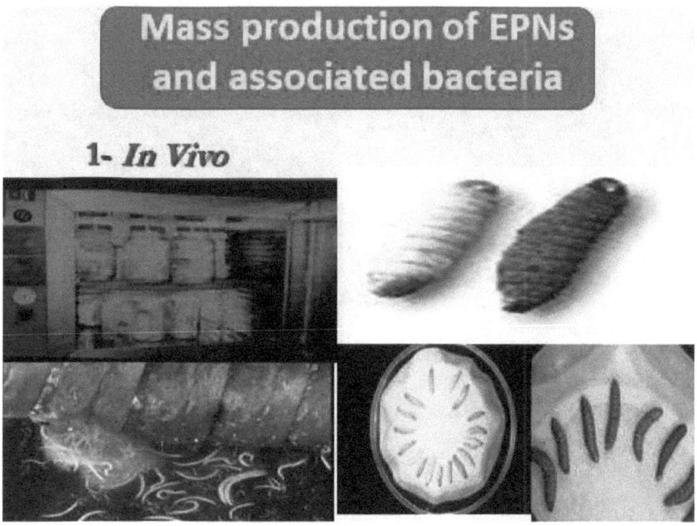

2- In vitro

Photorhabdus spp

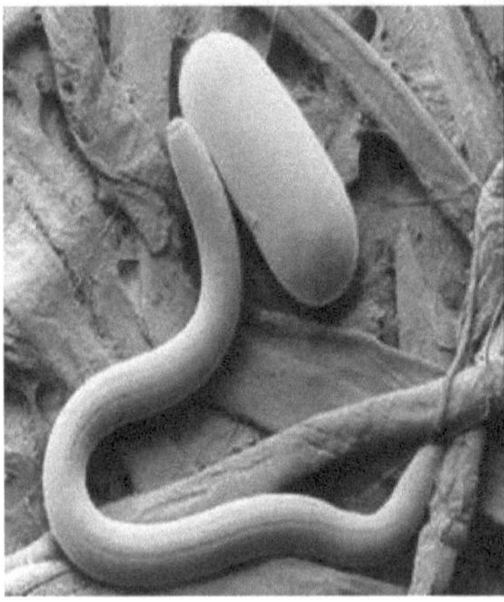

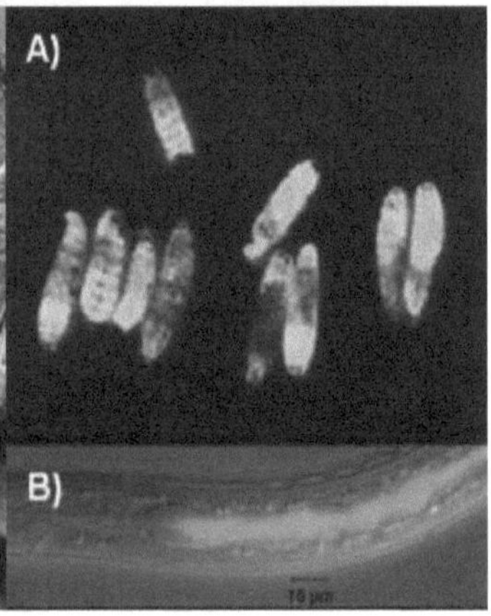

Xenorhabdus spp:

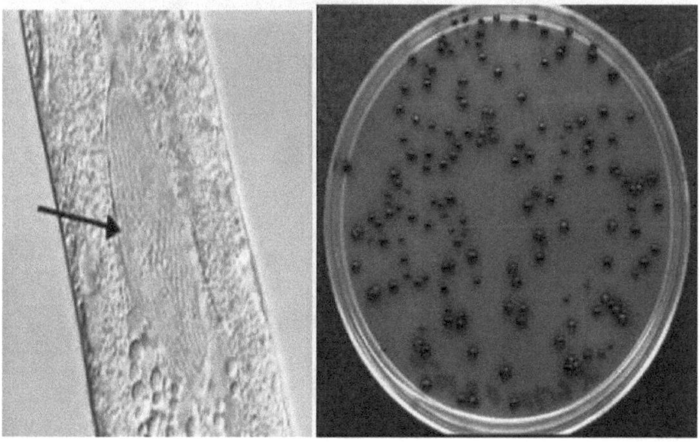

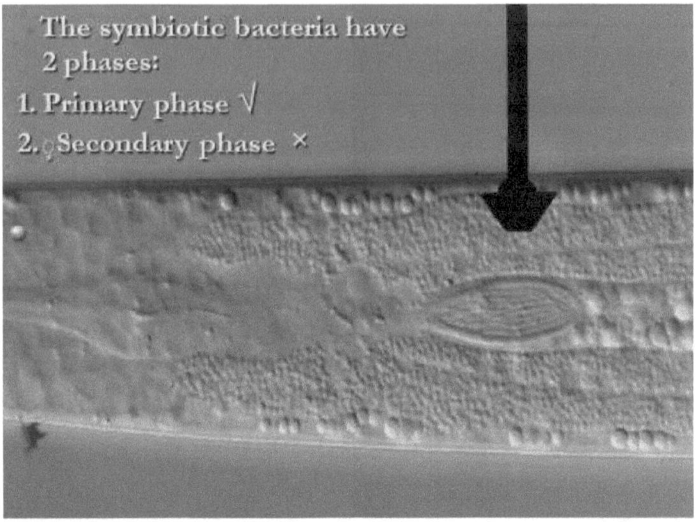

Phases Of Symbiotic Bacteria

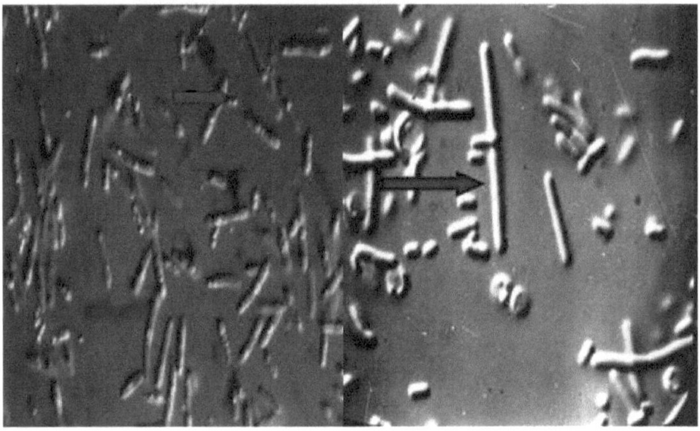

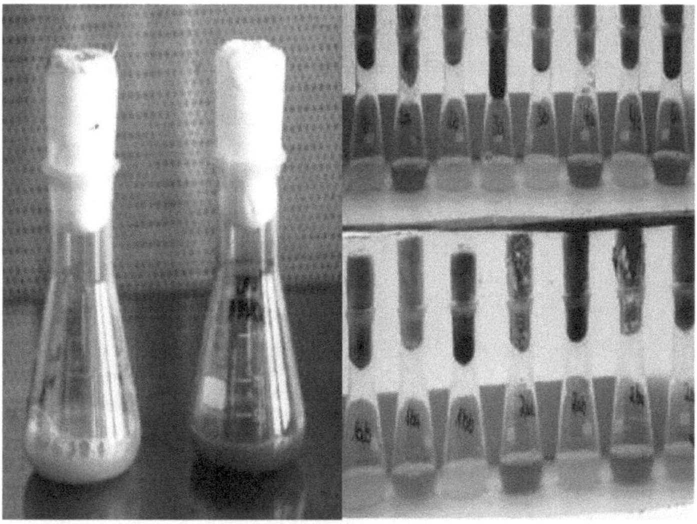

Foraging strategies

- The foraging strategies of entomopathogenic nematodes vary between species, influencing their soil depth distributions and host preferences. Infective juveniles use strategies to find hosts that vary from ambush and cruise foraging (Campbell 1997)
- In order to ambush prey, some *Steinernema* species nictate, or raise their bodies off the soil surface so they are better poised to attach to passing insects, which are much larger in size (Campbell and Gaugler 1993).
- Many *Steinernema* are able to jump by forming a loop with their bodies that creates stored energy which, when released, propels them through the air (Campbell and Kaya 2000).
- **Other species adopt a cruising strategy and rarely nictate. Instead, they roam through the soil searching for potential hosts.**
- These foraging strategies influence which hosts the nematodes infect.

- For example, ambush predators such as *Steinernema carpocapsae* infect more insects on the surface, while cruising predators like *Heterorhabditis bacteriophora* infect insects that live deep in the soil (Campbell and Gaugler 1993).

Application

- Although the biological control industry has acknowledged EPNs since the 1980s, relatively little is understood about their biology in natural and managed ecosystems (Georgis 2002).
- Nematode-host interactions are poorly understood, and more than half of the natural hosts for recognized Steinernema and Heterorhabditis species remain unknown (Akhurst and Smith 2002).
- Information is lacking because isolates of naturally infected hosts are rare, so native nematodes are often baited using *Galleria mellonella*, a lepidopteran that is highly susceptible to parasitic infection.
- Laboratory studies showing wide host ranges for EPNs were often overestimates, because in a laboratory, contact with a host is assured and environmental conditions are ideal; there are no "ecological barriers" to infection (Kaya and Gaugler 1993, Gaugler *et al.* 1997).
- Therefore, the broad host range initially predicted by assay results has not always translated into insecticidal success.
- The lack of knowledge about nematode ecology has resulted in unanticipated failures to control pests in the field. For example, parasitic nematodes were found to be completely ineffective against blackflies and mosquitoes due to their inability to swim (Lewis *et al.*1998).
- Efforts to control foliage-feeding pests with EPNs were equally unsuccessful, because nematodes are highly sensitive to UV light and desiccation (Lewis *et al.*1998).

- Comparing the life histories of nematodes and target pests can often explain such failures (Gaugler *et al.* 1997).
- Each nematode species has a unique array of characteristics, including different environmental tolerances, dispersal tendencies, and foraging behaviors (Lewis *et al.* 1998).
- Increased knowledge about the factors that influence EPN populations and the impacts they have in their communities will likely increase their efficacy as biological control agents

Using Entomopathogenic Nematodes for Crop Insect Pest Control

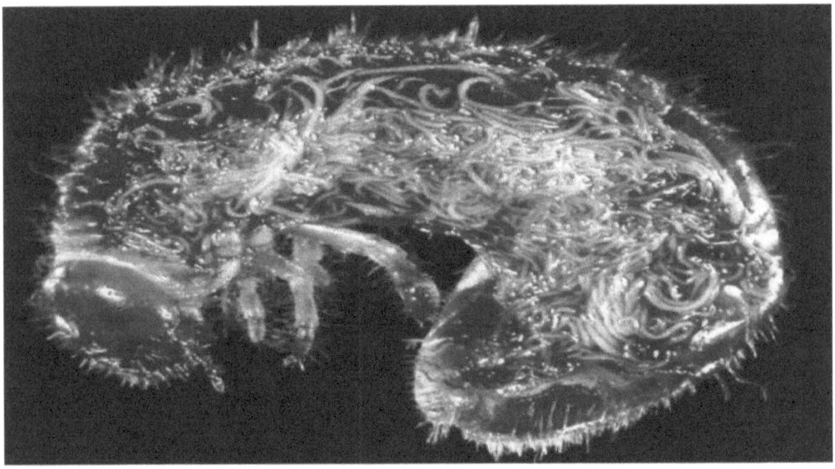

How Entomopathogenic Nematodes Control Insect Pests

- An understanding of host-finding strategies will help you properly match entomopathogenic nematode species to pest insects to ensure infection and control (Gaugler 1999).
- Only entomopathogenic nematodes in the infective juvenile stage will survive in the soil and find and penetrate insect pests.
- Infective juvenile entomopathogenic nematodes locate their hosts in soil Ambushing.
- Entomopathogenic nematodes that use the ambushing strategy tend to remain stationary at or near the soil surface and locate host insects by direct contact (Campbell et al. 1996).
- An ambusher searches by standing on its tail so that most of its body is in the air, referred to as "nictation." The nictating nematode attaches to and attacks passing insect hosts.
- Ambusher entomopathogenic nematodes most effectively control insect pests that are highly mobile at the soil surface, such as cutworms, armyworms, and mole crickets.
- Cruising. Entomopathogenic nematodes that use the cruising strategy are highly mobile and able to move throughout the soil profile. Cruisers locate their host by sensing carbon dioxide or other volatiles released by the host.
- Cruiser entomopathogenic nematodes are most effective against sedentary and slow-moving insect pests at various soil depths, such as white grubs and root weevils.

Infection

- Generally, several entomopathogenic nematodes will infect a single insect host. Infective juvenile nematodes penetrate the insect's body cavity either through natural body openings (such as the mouth, anus, genital pore, or breathing pore; Figure 4) or by breaking the outer cuticle of the insect; heterorhabditids do this using a dorsal "tooth" or hook.
- Once inside the body cavity of the host, the infective juveniles release bacteria that live symbiotically within the entomopathogenic nematode's gut but do not harm the nematode.
- The nematode-bacterium relationship is highly specific: only *Xenorhabdus spp. bacteria co-exist with steinernematids, and only Photorhabdus bacteria co-exist with heterorhabditids.* Once released into the host, the bacteria multiply quickly and under optimal conditions cause the host to die within 24 to 48 hours.
- Entomopathogenic nematodes feed on both bacteria they release and host insect tissue. After a few days inside the host, entomopathogenic nematodes mature to the adult stage.
- These adult entomopathogenic nematodes produce hundreds of thousands of new juveniles that may undergo several life cycles within a single host

Infective juvenile entomopathogenic nematodes searching for an entry portal into a host (photo provided by Randy Gaugler)

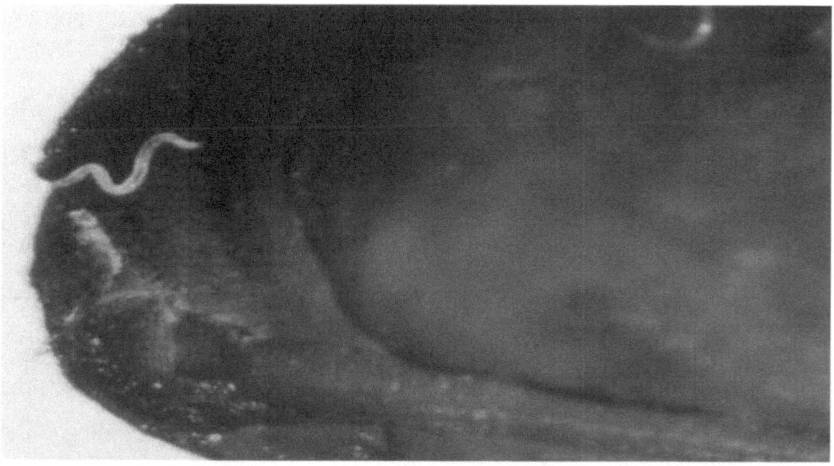

Hundreds of thousands of new entomopathogenic nematodes within a host insect (photo provided by Randy Gaugler).

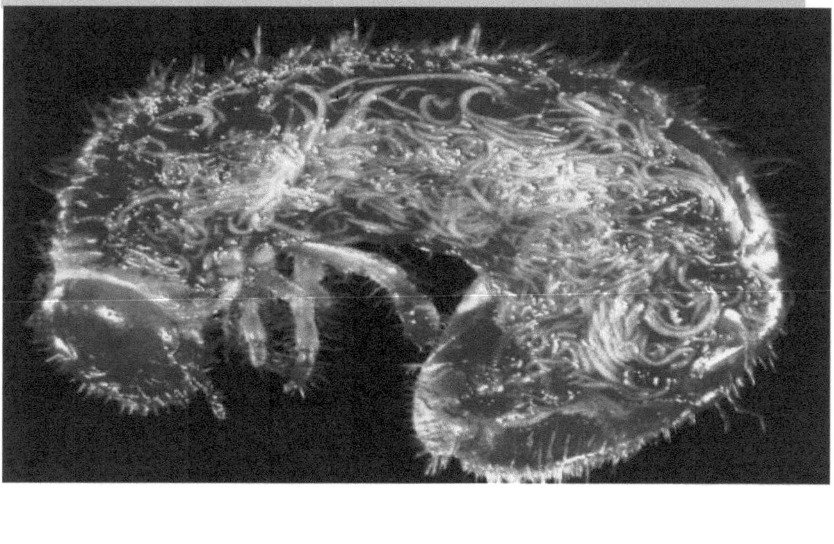

- When the host has been consumed, the infective juveniles, armed with a fresh supply of bacteria, emerge from the empty shell of the host, move into the soil, and begin the search for a new host.

- A protective exterior cuticle surrounds the infective juvenile, protecting it from the environment and predators.
- Under ideal conditions, steinernematids emerge 6–11 days after initial infection and heterorhabditids emerge 12–14 days after initial infection (Kaya and Koppenhfer 1999).
- The duration of infective juvenile survival in soil is unknown because they can become prey to invertebrates and microorganisms.
- particular pest more effectively than another species. Therefore, the insect pest must be identified before choosing the entomopathogenic nematode species most appropriate for biological control.

Using Entomopathogenic Nematodes

- Worldwide, over 80 species of entomopathogenic nematodes have been identified and 11 commercialized (Kaya and Koppenhfer 1999).
- The different species of entomopathogenic nematodes vary in the range of insects they attack, environmental needs, and stability in commercial products (Gaugler 1999).
- A given species of entomopathogenic nematode may also control a particular pest more effectively than another species.
- Therefore, the insect pest must be identified before choosing the entomopathogenic nematode species most appropriate for biological control.

Shelf Life

- In general, entomopathogenic nematodes do not have a long shelf life. Many microbial insecticides, including *Bacillus thuringiensis,* have a resting stage facilitating long-term storage.
- The infective juvenile entomopathogenic nematode stage is not a resting stage; juveniles are metabolically active and use energy reserves while in formulation (Lewis 1999).
- For this reason, it is advisable to order entomopathogenic nematodes only 3–4 days prior to application. Entomopathogenic nematodes should be shipped by overnight delivery in their infective juvenile stage and used within 1–2 days after arrival.
- Examine the entomopathogenic nematodes upon receipt to make sure they arrived alive. The shipment container should not feel warm or hot. Open the container and check the color and odor of the nematodes
- To the naked eye, the nematodes on a sponge formulation will appear as a light tan or gray paste, while nematodes in vermiculite or liquid suspension will not be discernible from the carrier material.
- The container should have a mild odor; if there is a strong smell, like ammonia, then it is likely the nematodes are dead.
- If the formulation is a sponge or vermiculite, remove a tiny portion of the product with tweezers and place in a teaspoon of cool water (approximately 60°F) for six hours.
- If the formulation is a liquid suspension, swirl the liquid to ensure distribution of the nematodes and remove a small droplet (about 0.05 ml).

- Entomopathogenic nematodes should be stored in their shipment containers under refrigeration until ready for use.
- The storage life of entomopathogenic nematodes is species- and formulation-dependent. Specific storage instructions will be included with the entomopathogenic nematode shipment and should be carefully followed.

Soil Conditions

Entomopathogenic nematodes can die if they are applied to soils that are too dry, too hot, or too cold, or if they are exposed to ultraviolet (UV) light from the sun.

Nematodes live in the water-filled spaces, or pores, between soil particles.

They need water to move and successfully locate a host, and oxygen to survive. Heavy clay soils hold water well, but may contain too little oxygen, and the small pore space may restrict nematode movement.

Sandy soils must be irrigated to maintain the water-filled pores

If applying entomopathogenic nematodes after an extended dry period, break the crust of the dry soil with a rake or harrow and irrigate the soil before the application, preferably to a depth of 4–6 inches. Remove plant debris before the application, as it will prevent the nematodes from reaching the soil surface.

Soil temperatures between 77°F and 82°F are ideal for applying all entomopathogenic nematode species. In general, soil temperatures greater than 85°F can decrease the efficacy of some nematode species, while soil temperatures less than 50°F can immobilize others at the soil surface, causing them to be exposed to UV light that can kill them.

The range of soil temperatures that nematode species can survive and infect host insects does vary, however. For example, *S. feltiae* can be effective at 57°F, while *S. riobrave* can be effective at 95°F.

Entomopathogenic nematodes should be applied late in the day or on a cool, overcast day when light and temperatures are low.

Apply them to moist soil following either a rainfall or irrigation, and lightly irrigate afterward.

This washes them into the soil and decreases soil surface temperatures.

Do not over-irrigate, as saturated soil will impede nematode activity due to lack of oxygen.

Preparing for Application

Entomopathogenic nematodes should be prepared for field application no earlier than one hour ahead of time.

If nematodes are in a liquid suspension, shake the shipment container well and pour the liquid into the application container (e.g., tank, backpack sprayer, or watering can).

Rinse the shipment container twice with cool water (approximately 60°F), and pour the rinse water into the application container.

If nematodes are on a sponge, soak the sponge in one gallon of cool water for 10 minutes, then pour the water into the application container.

Rinse the sponge several times, pouring the rinse water into the application container after each rinse. If nematodes are in vermiculite, add the vermiculite-nematode mixture directly to water in the application container and stir until dispersed.

Once the nematodes have been mixed with water, agitate the mixture every five minutes to keep the nematodes in suspension and supplied with oxygen.

Pesticide Safety

Entomopathogenic nematodes, like all biological control organisms, are considered pesticides. How ever, entomopathogenic nematodes are exempted from federal pesticide registration requirements under 40CFR152.20 (a) because they belong to a group of specific biocontrol agents.

Questions regarding pesticide registration should be directed to your state's Department of Agriculture Pesticide Division.

Nemátodes

Steinernema feltiae

effectiv against
Thrips-pupae in the soil.

Viruses

A. Baculovirus is sprayed onto foliage

B. Caterpillar consumes the virus

C. The protein encapsulating the Baculovirus DNA dissolves and the DNA enters the stomach cells

D. Baculovirus DNA is replicated by the stomach cells until the stomach cells rupture

E. The caterpillar stops eating

F. Baculovirus is spread throughout the caterpillar causing a general systemic infection

G. The caterpillar dies within days

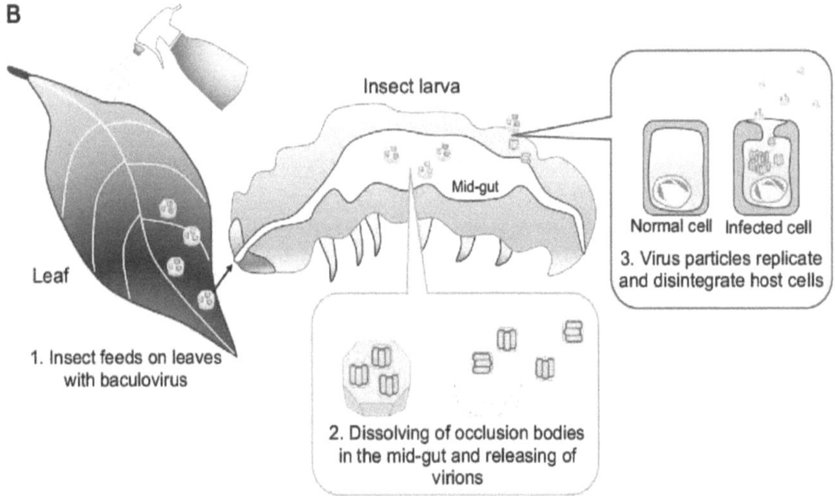

UGA1323075

Predators

- Insect, mite or spider that attacks & feeds on its prey

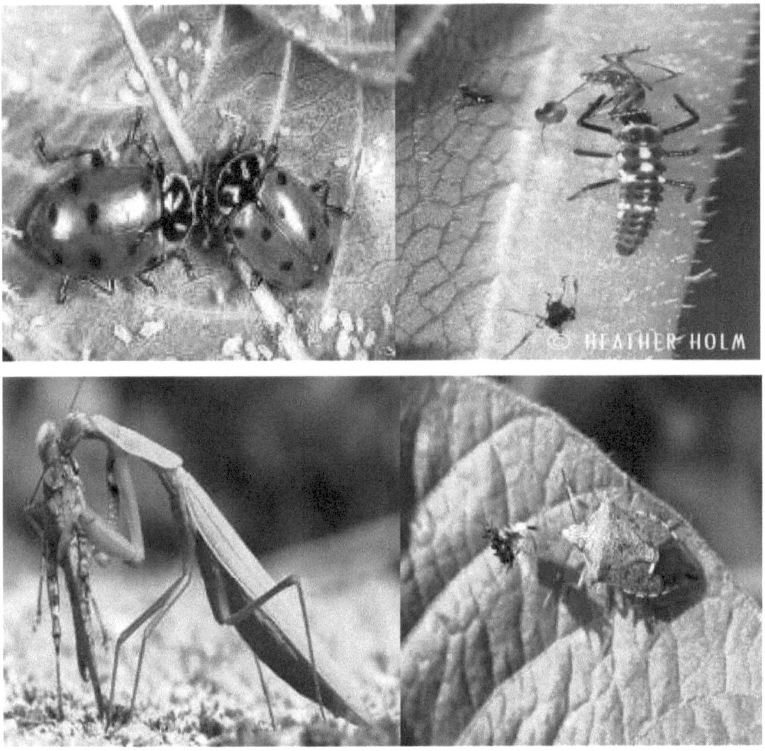

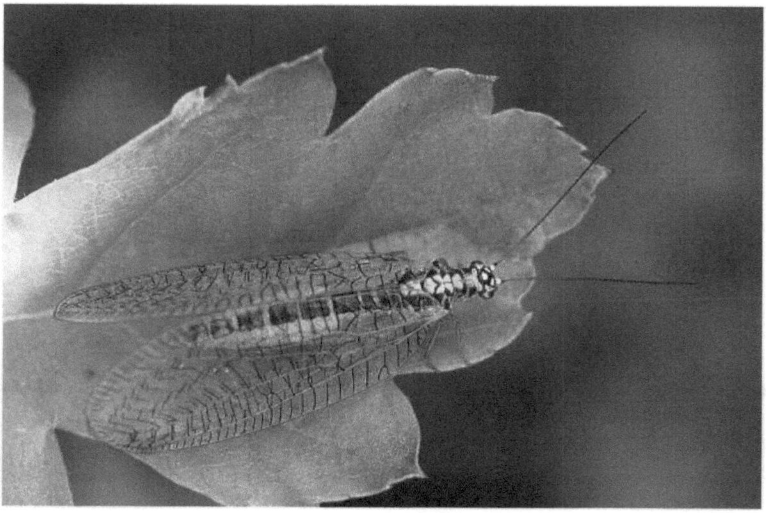

Chrysoperla carnea

(Fam: Chrysopidae)

Order: Neuroptera

The description

A delicate insect with a body about 1 cm long. The general color is light green - the veins of the wings are reticular

Economic importance:

Larvae are predatory for aphids, whiteflies, thrips and some types of scale insects

Larva of Chrysoperla carnea

Egg of Chrysoperla carnea

Chrysoperla carnea larva preys on aphids

Ceuta varigata

(Fam: Myrmeleonidae)

The description:

Its length is about 1.5 cm. The general color is yellowish-gray - the veins of the wings are reticular

Economic importance:

The larvae are predators of ants

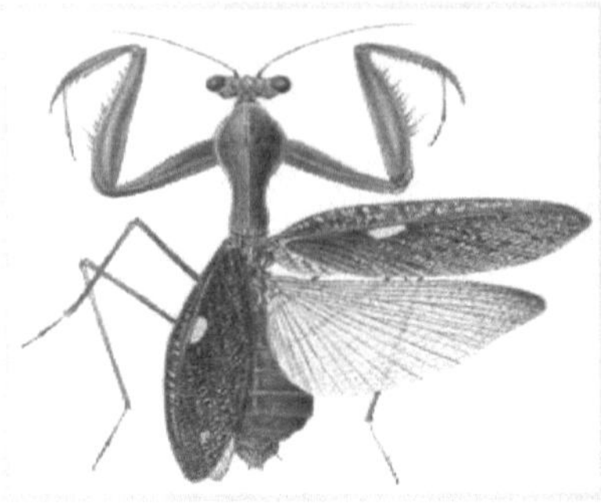 

Sphodromantis bioculata

Order: Dictyoptera

(Fam: Mantidae)

The description:

Its length is about 4 cm. The general color is green. The front wing is leathery. The hind wing is membranous. The front legs are modified into sniper legs.

 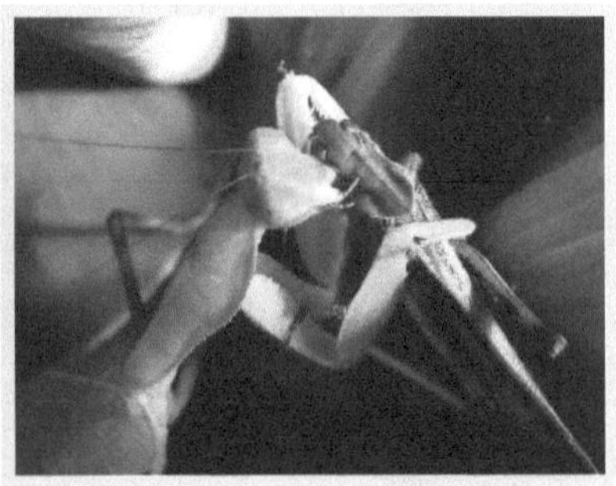

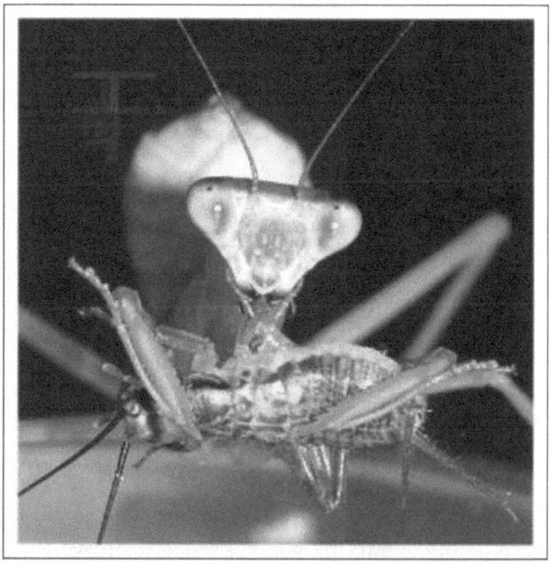

Economic importance:

Predator of flies, hoppers, spiders, ants and jassids

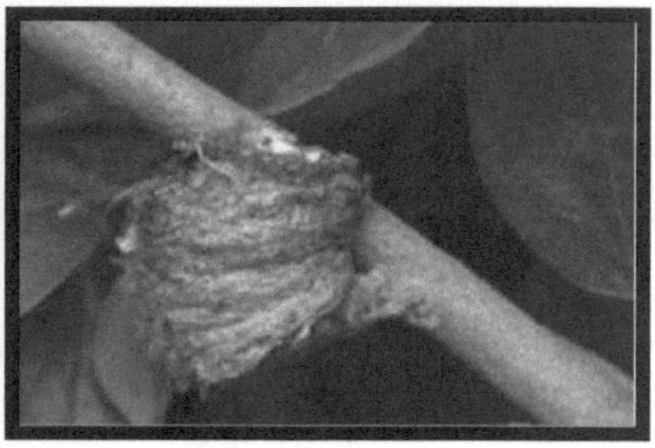

Egg sack

Labidura riparia

Order:Dermaptera

(Fam: Labiduridae)

The description:

Its length is about 2 cm, the general color is brown - the anal horn is clipped - the front wing is short cutaneous - the hind wing is semicircular membranous

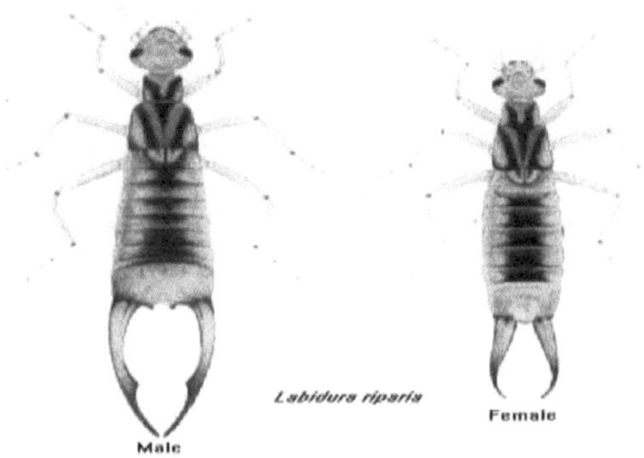

Economic importance:

Predator of Lepidoptera larvae as well as larvae of beetles and most species of aphids

Crocthemis erythraea

Order: Odonata

(Fam: Libellulidae)

Economic importance:

Nymphs prey on aquatic insects such as larvae and pupae of mosquitoes, and entire insects prey on what they encounter in flight, such as flies and mosquitoes.

Paederus alfierii

Order:Coleoptera

(Fam: Staphylinidae)

The description

Its length is about a name - the sensor horn is threadlike - the body is spindle-shaped - the color of the head and the sheaths and the end of the abdomen are bright blue and the rest of the body is orange and the sheaths are short

Coccinella undecium punctata

(Fam: occinellidae)

The description

Its length is about 4 mm - the general color is yellow-orange - there are 11 black points on the sheaths.

Larva of ***Coccinella undecium punctata***

Economic importance

The adults and their larvae prey on aphids, mealybugs and newly hatched eggs and larvae of lepidopterans.

Rodalia cardinalis

(Fam: Coccinellidae)

The description

Its length is about 4 mm - the general color is honey red - extends in the middle of the two sheaths with a zigzag black stripe and on each sheath there are two black spots.

Economic importance

The whole bug and its larvae prey on the Australian mealybug

Orius niger

Order:- Hemiptera

(Fam:Anthocoridae)

The description

The length of the insect is about 7 mm - the general color is brown - the front wing is semi-colon

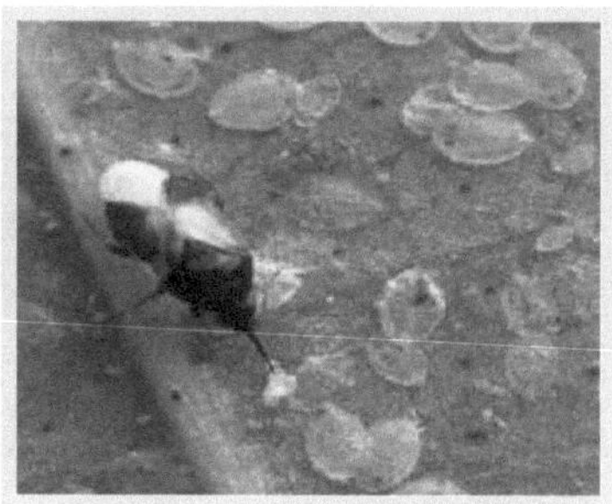

Economic importance

The nymph and the adult insect prey on the eggs of the cotton leaf worm and prey on aphids - thrips - white flies - dreaming -

and it is considered one of the most important predators in the Egyptian fields.

Syrphus corolla

Order:- Diptera

(Fam: Syrphidae)

The description

Its length is about 1 cm - the general color is black - there are yellow stripes on the abdomen.

Hosts: predators of aphids

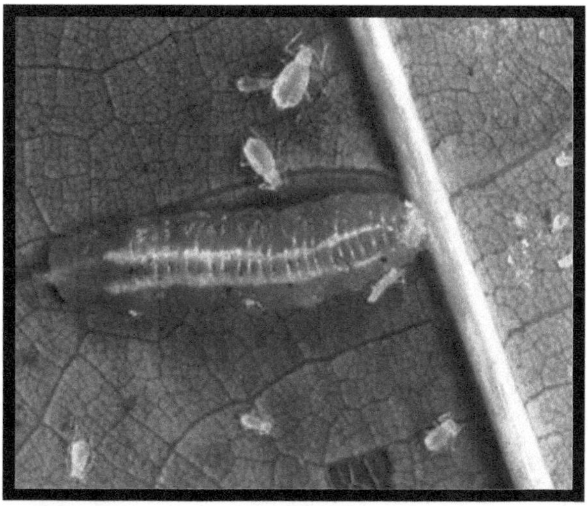

Economic importance

The larvae prey on aphids, scale insects, and mealybugs.

Parasites

Parasites are external parasitism

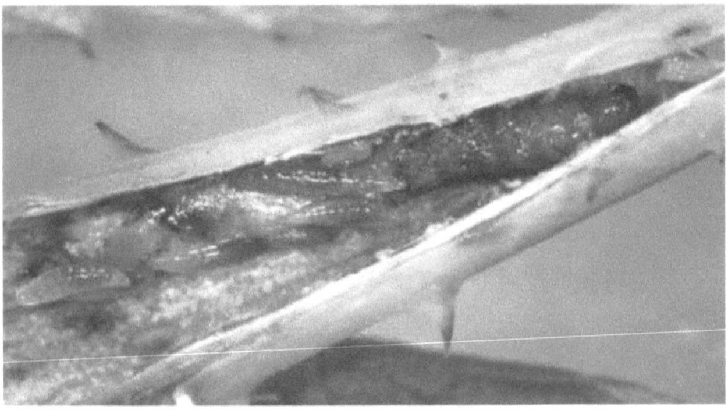

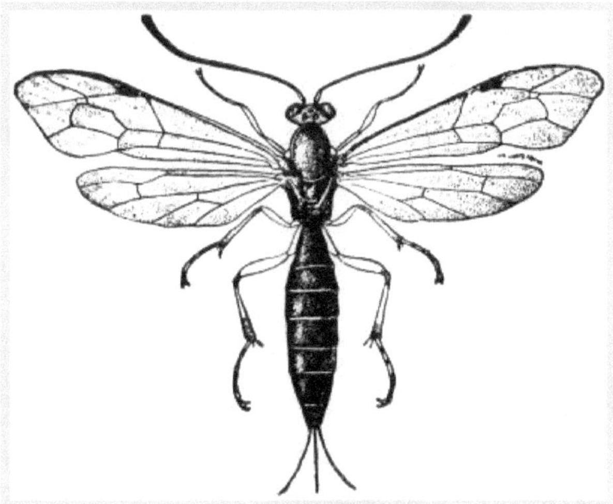

Pimpla roborator

Order:- Hymenoptera

(Fam: Ichneumonidae)

The description:

The length of the female is about 2.5 cm, including the egg-laying machine - the color of the head, chest and antennae is black - the abdomen is reddish-brown on the front edge of the front wing its eyes or stigma

Economic importance

An ectoparasite on the larvae of the pink almondworm and the larvae of the eggplant stemworm

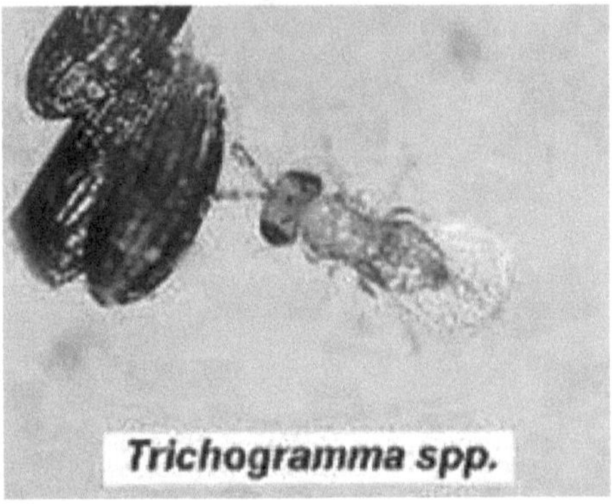

Trichogramma evanescens

(Fam: Trichogrammatidae)

The description:-

The length of the insect is about .5 mm - its general color is yellow - the color of the eyes is red

Economic importance:

A true egg parasite that parasitizes the eggs of most insects of the squamous wing order

A
B
C
D

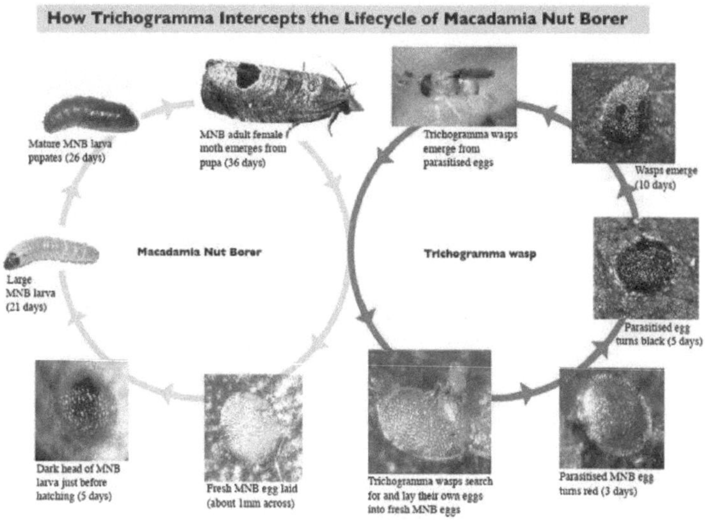
How Trichogramma Intercepts the Lifecycle of Macadamia Nut Borer
Mature MNB larva pupates (26 days)
MNB adult female moth emerges from pupa (36 days)
Trichogramma wasps emerge from parasitised eggs
Wasps emerge (10 days)
Macadamia Nut Borer
Trichogramma wasp
Large MNB larva (21 days)
Parasitised egg turns black (5 days)
Dark head of MNB larva just before hatching (5 days)
Fresh MNB egg laid (about 1mm across)
Trichogramma wasps search for and lay their own eggs into fresh MNB eggs
Parasitised MNB egg turns red (3 days)

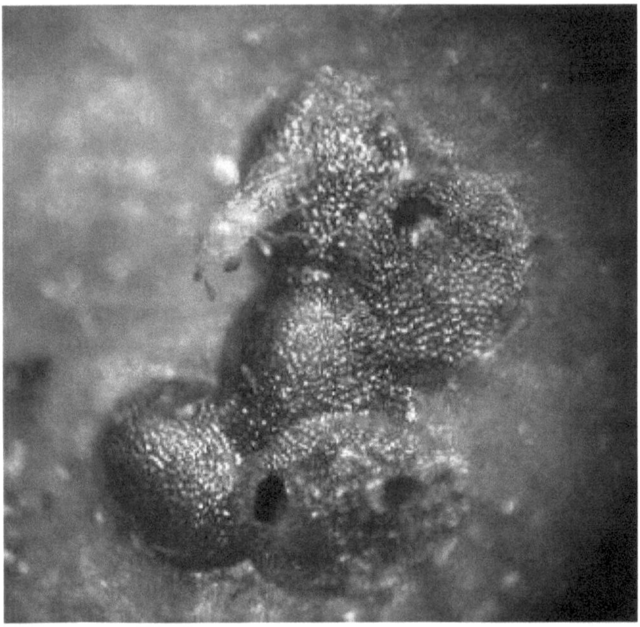

Bracon hebetor

Bracon hebetor, is an ectoparasitoid of the larvae of many pyralid pest-species, that attack stored grains. The parasitoid is considered to have a potential for the biological control, of many other lepidopterous pests.
The female injects venom that induces complete paralysis of host.
The female, placing several eggs on the host.

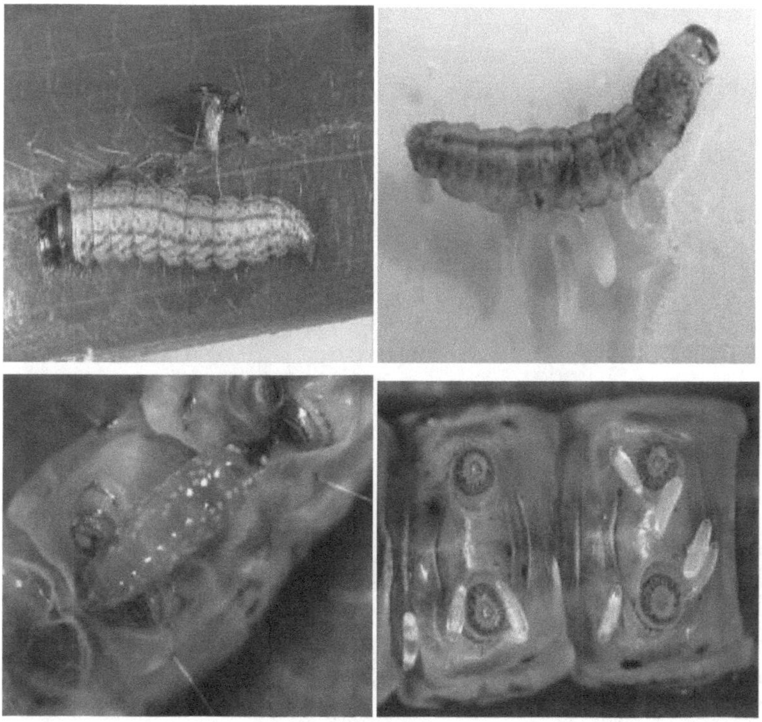

Mass rearing of *Bracon sp.*

The adults of *B. hebetor*, were reared on a large scale, by using the larvae of the greater wax moth, *Galleria mellonella*

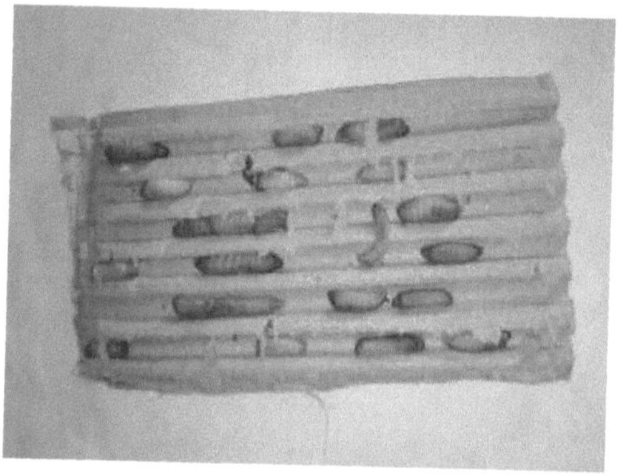

Brachymeria femorata

(Fam: Chalcididae)

The description:-

The length of the insect is about 1 cm, the thigh of the hind leg is enlarged, and it bears on its edge several teeth - the color of the legs is yellow

Economic importance:

Internal parasite of cabbage flour pupae

Aphytis holoxanthus

(Fam: Aphelinidae)

The description:-

The length of the insect is about 1 mm, the general color is yellow - the egg laying machine is clear

Economic importance:

An ectoparasite on black scale insects that parasitizes on the nymph of a second age in the case of females and on the pupal stage and before the pupa in the case of the male

Tachina larvarum

Order:- Diptera

(Fam: Tachinidae)

The description:-

The length of the insect is about 1 cm. The general color is black - on the dorsal surface of the abdomen there are 3 transverse areas of light gray color and on the abdomen long hair

Economic importance:

An internal parasite on the larvae of the cotton leaf worm and the green worm

Gonia capitata

(Fam: Tachinidae)

The description:-

The length of the insect is about 1 cm. The general color is gray - there are two spots on the abdomen, their color is yellow, and on the abdomen is long hair

Economic importance:

Internal parasite on cutworm larvae

Advantages of biological control

1. Effective against insects which are resistant to pesticides.
2. Biological control is a very specific strategy
3. Cheap in the long run.
4. Safety to the environment.

Disadvantages of Biological Control

1. It's a slow process. It takes a lot of time and patience.
2. The introduced species may have unknown ecological effects, and even become a pest itself. e.g. the cane toad and the harlequin ladybird.
3. Frequent input is needed to maintain a population balance between predator and prey.

Approaches to Biological Control

- There are three general approaches to biological control;
- ➢ Importation
- ➢ Augmentation
- ➢ Conservation.
- ➢ Each of these techniques can be used either alone or in combination in a biological control program.

• Importation

- This is a process which requires extensive research into the biology of the pest, potential natural enemies and their biology.
- After suitable natural enemies are found and studied they must undergo quarantine.

- Although this process is long and complex, when it is successful, the results can be impressive and permanent, as long as care is taken in production practices to minimize negative effects on the natural enemy

• Conservation

Conservation of natural enemies involves either, reducing factors which interfere with natural enemies or providing resources that natural enemies need in their environment.

Many factors can interfere with the effectiveness of a natural enemy.

1) Reducing pesticide use:Pesticide applications may directly kill natural enemies.
2) Providing habitat and resources for natural enemies.To be effective, natural enemies may need access to; alternate hosts, adult food resources, overwintering habitats, constant food supply

Augmentation

Augmentation is the direct manipulation of natural enemies to increase their effectiveness. in which natural enemies are produced then released

Figure 3. An adult of *Orius laevigatus* (Fieber) feeding on artificial diet (photo Y. Arijs).

Fundamentals of successful biological control

1- Correct identification of the types of target pests and their natural enemies
2- Good knowledge of the biological and environmental aspects of the pest and its vital enemies
3- Determining and knowing the critical economic limit of the pest
4- Reducing the use of chemical pesticides and relying on agricultural control
5- Choosing the appropriate biological enemy for the pest
6- Proper selection The method and timing of launching natural enemies
7- Accurate evaluation of the results of the launch to judge the success of the operation or not

Breeding the biological enemy on insect families

Mass rearing of beneficial insects

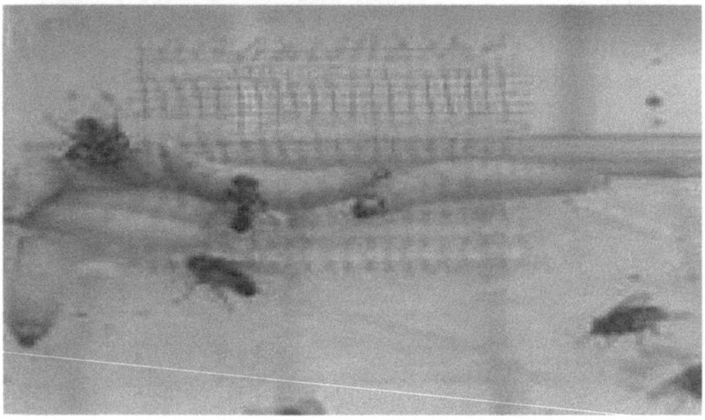

Figure 3. An adult of *Orius laevigatus* (Fieber) feeding on artificial diet (photo Y. Arijs).

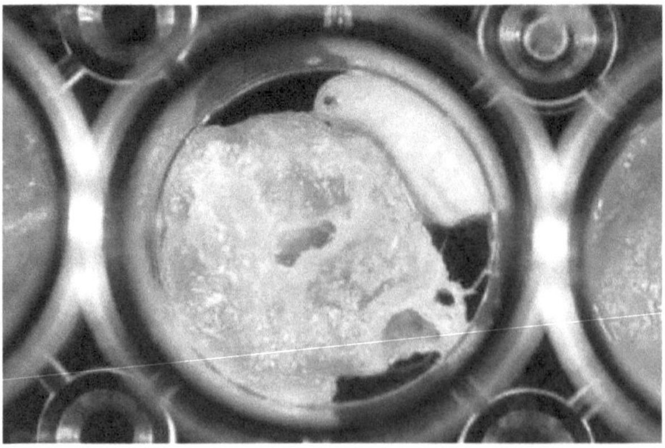

Attractants, repellents, nutritional inhibitors, hormones and plant extracts

Modern methods of control look at the use of such chemicals or their analogues.

Manufactures can exploit the pest community by releasing the insect's behavior.

Or causing confusion in the physiological processes of insects

Attractive materials

Attractive materials give directional instructions so that the insect can move from Long distances to the smell area.

They can follow smells, search for food sources, places to lay eggs, or to mate.

Pheromones

Pheromones are chemicals or mixture of chemicals released by organism to outside to communicate with each other in the same species. Insects send these chemical signals to help attract mates, warn others of predators, or find food.

Types of pheromones in insects

1- Sexual pheromones: lead to the attraction of males to females, or vice versa.

2- Aggregation pheromones: as in migratory locusts.

3- Alarm pheromones: alert individuals to defend the hive, like bees.

4- Tracking pheromones: for the purpose of insects, especially social ones.

5- Ripening pheromones: present in migratory locusts, secreted in non-pigmented phases mature to motivate him to prepare for migration and maturity in one time

Sex Pheromones

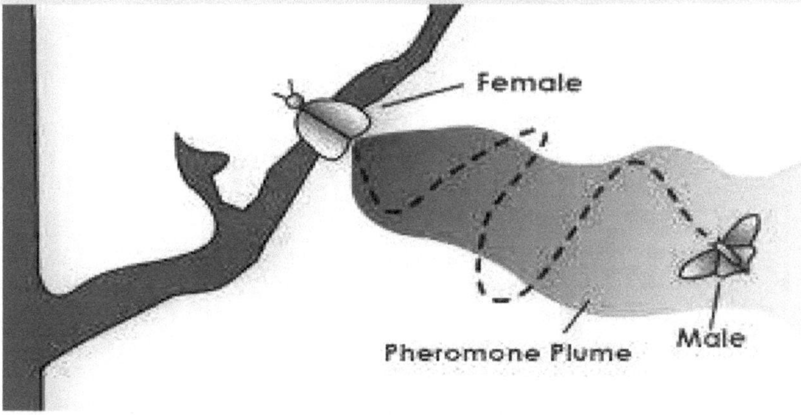

Alarm Pheromones

Hornet wasp

Honey bee

Trail pheromones

Found mostly in social insects, including ants, termites, bees and wasps. When a worker locates a resource, she lays down a trail when returning to the colony that other workers can use to find the resource. Flying insects use trail pheromones to stimulate colony members to enter the hive.

Pheromone Traps

Using specific pheromones, traps can be used to monitor target pests in agriculture or in residential areas. By constantly monitoring for insects, it may be possible to detect an infestation before it occurs.

Early detection of pest insects using pheromone traps can also lessen damage to agriculture and other plants.

Allelochemicals

➢ **1- Kairomones**

➢ **Kairomones→→** benefit the receiver but disadvantage the producer .

- For example, fifth instar larvae of *Ephestia kuehniella* produce secretion by Mandibular glands acts as kairomone for the larval ectoparasitoid *Bracton hebetor.*

➢

➢ **Allelochemicals**

➢ **2- Allomones**

➢ **Allomones→→**benefit the producer by modifying the behavior of the receiver although having a neutral effect on the receiver.

◆ For example, defensive and/or repellent chemicals are allomones that advertise distastefulness and protect the producer from lethal experiment by prospective predators.

The effect on a potential predator is considered to be neutral, as it is warned from wasting energy in seeking a distasteful meal.

Allelochemicals

3- Synomones

- **Synomones** →→benefit both the producer and the receiver.

For example *The floral scent that attract pollinators

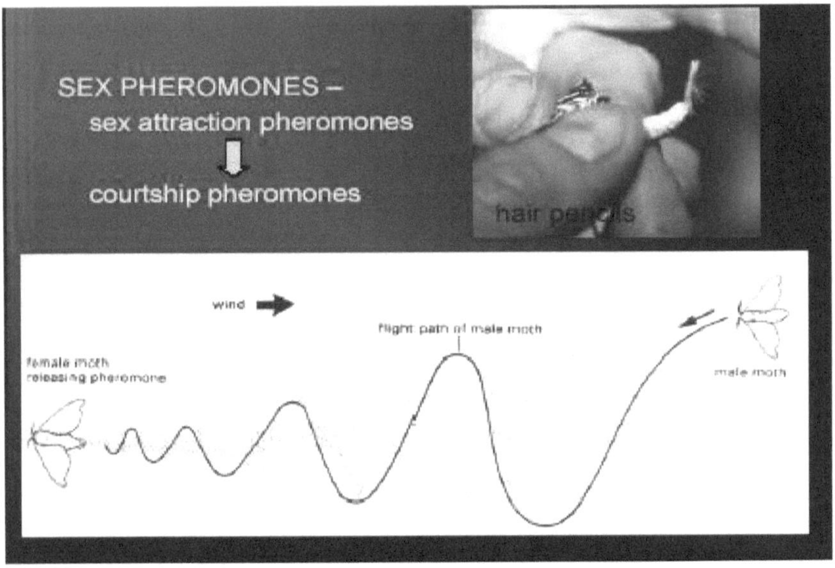

Pheromone Disperser Examples

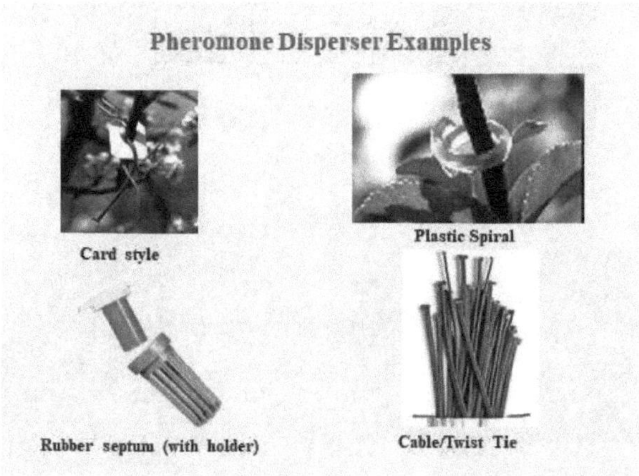

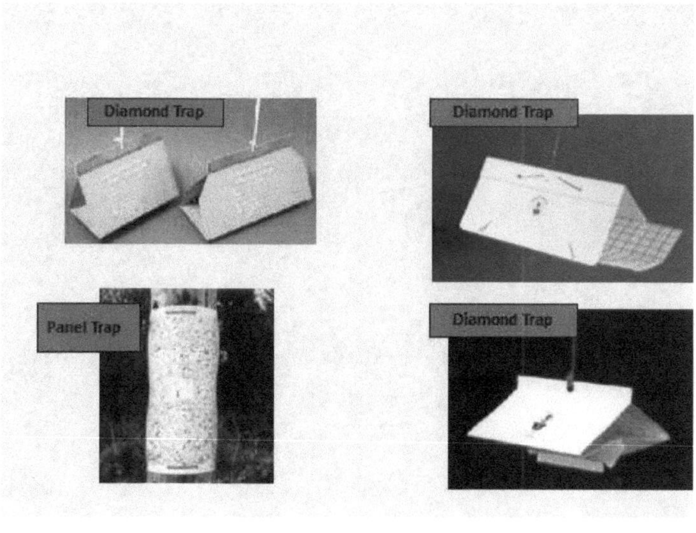

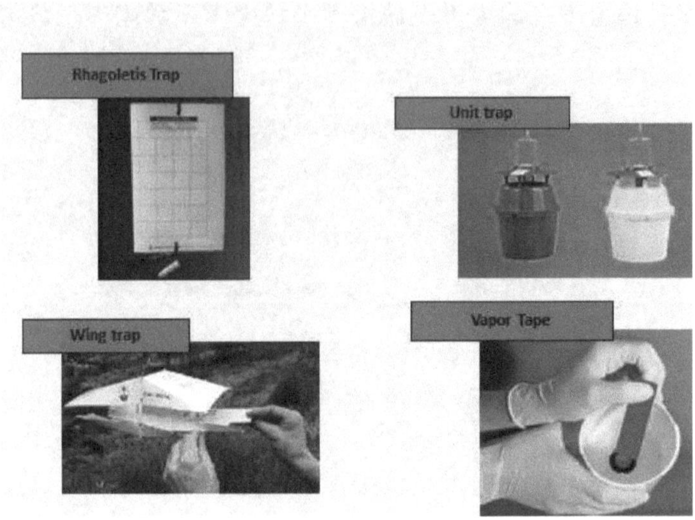
Rhagoletis Trap
Unit trap
Wing trap
Vapor Tape

Factors affecting the efficiency of pheromone traps

1- (Insect density): The higher the number of insects, the greater the number of insects attracted to the traps.

2- (infestation site): the closer the traps are to the site of the infestation, the more effective the traps are in attracting insects.

3- (Number of traps used): The number of traps should be proportional to the area and the rate of infection

4- (Design of traps): It is necessary to choose the appropriate design of the trap in terms of shape according to the conditions of the affected place.

5- (Pheromone concentration): The pheromone concentration and the rate at which it is released from its packages must be appropriate.

6- (Knowledge of insect behavior): Knowledge of insect behavior facilitates choosing the most appropriate place and time to place traps.

Some uses of sex pheromones traps

1- Detection of the presence of various insects.

2- Determine the location of the injury.

3- Estimation of pest density

4- Determining the appropriate control method and its effectiveness.

Repellent materials

Chemicals that prevent harm to plants, animals, or materials such as fabrics and wood by making them unattractive or unpalatable.

Such as naphthalene, which repels clothes and carpet moths.

And when repellents are sprayed on the vegetative group of plants, they provide coverage complete for him

The oils were used to repel legume beetles and prevent them from laying eggs on legumes.

Nutritional supplements

Substances that stop eating permanently or temporarily and may cause the insect to starve

You cannot eat plants sprayed with these substances.

Hormones

It is an internal secretion that regulates a wide range of processes inside the insect's body, including growth, development and maturity.

Botanical Insecticides

Plants produce a vast and diverse assortment of compounds and the majority of them do not participate directly in growth and development. These phytochemicals, traditionally referred to as **secondary metabolites.**

Many of these compounds have been shown to have important adaptive in protection against herbivory

Repellents

They drive away the insect pest from the treated materials by stimulating olfactory or other receptors.

- Minimise pesticide residue
- Ensure safety of the people, food, and environment .

- For example, the essential oil of *Artemisia annua* was found as repellent against *Tribolium castaneum* and *Callosobruchus maculate*

Antifeedants/Feeding Deterrents

- Inhibit feeding or disrupt insect feeding by rendering the treated materials unattractive or unpalatable.
- Essential oil constituents such as thymol, citronellal and α-terpineol are effective as feeding deterrent against tobacco cutworm, *Spodoptera litura* synergism.

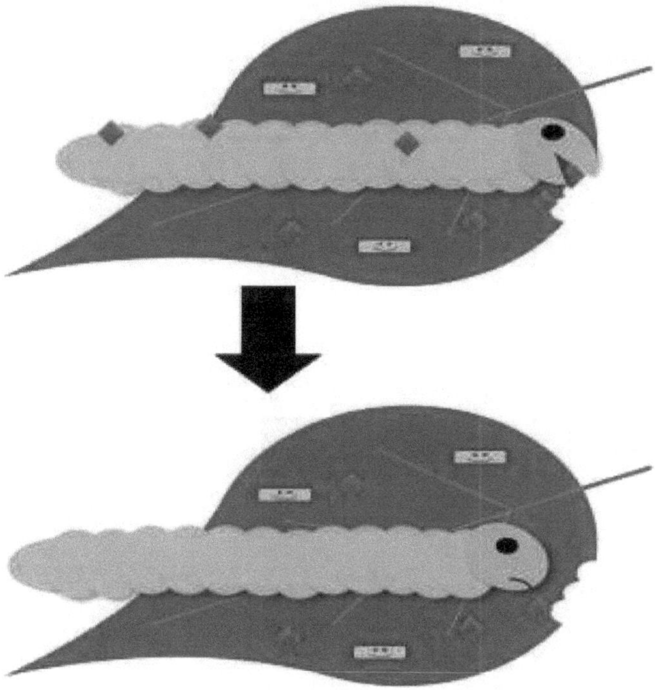

Insect Growth and Development Inhibitors

- Plant extracts showed deleterious effect on the growth and development of insects and reduced larval pupal and adult weight significantly, lengthened the larval and pupal periods, and reduced pupal recovery and adult eclosion

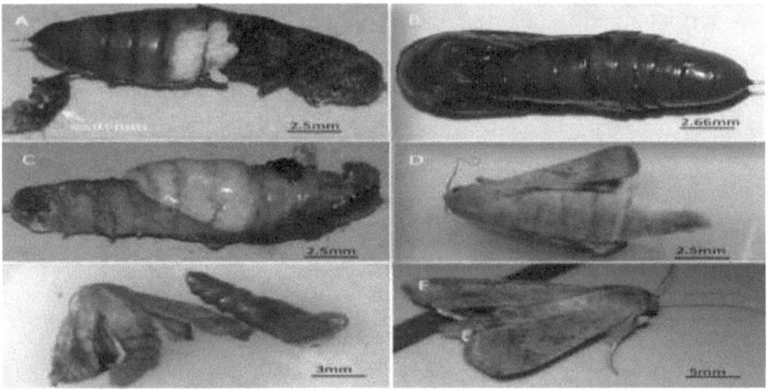

Plant extracts

Leads to:

- kill
- kicked out

The growth of insects in their different stages

Antifeedants

They are chemicals that prevent the start or continuation of feeding the insect on the appropriate host, and it does not matter that these substances have a repellent or toxic effect

Feed inhibitors are one of the modern trends in pest control, and they differ from insecticides in that they do not lead to direct killing of the pest or to expel it, while their effect is due to their ability to prevent feeding the pest and thus the insect dies of starvation if it does not find another host. Example: neem tree extracts (Zenzalkhat)

1. Nimbecidine EC:

0.03% Azadirachtin, Formulated by T. Stanes & Company Limited, India was used at concentration 4ml/L.

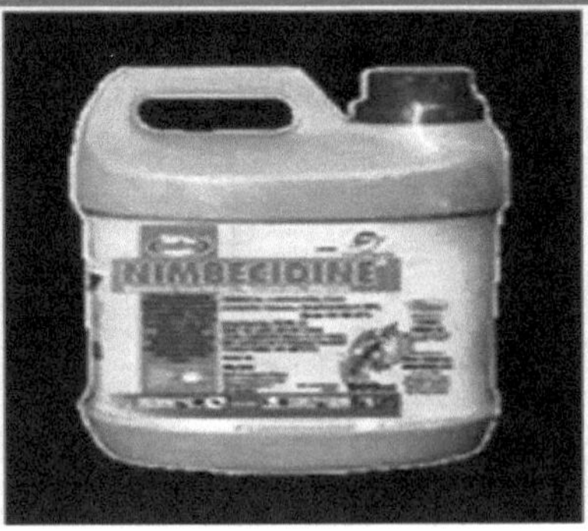

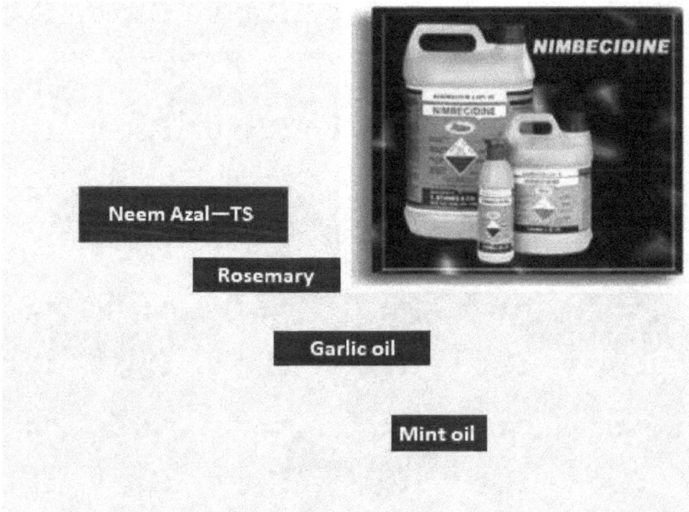

The repellent effect of these extracts increased with the increase in the number of sprays to five sprays compared to one or two sprays

Neem Azal—TS

⇩

Rosemary

⇩

Garlic oil

⇩

Mint oil

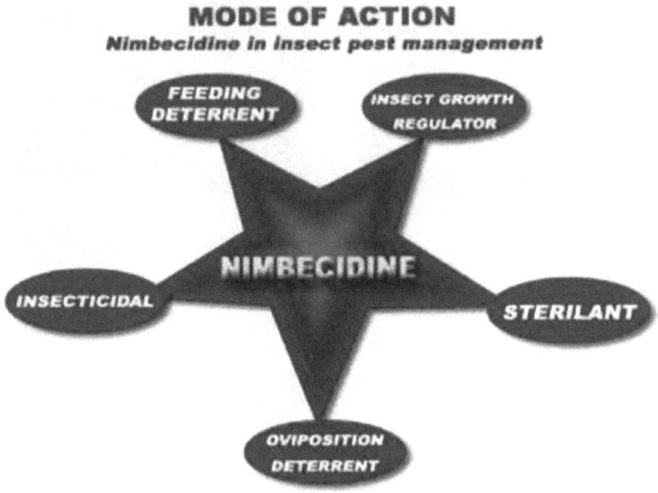

Mode of Action

Nimbecidine exhibits multiple modes of action. It acts as an:

- **Antifeedant**
- **Repellent**
- **Ovi-position Deterrent**
- **Insect Growth Regulator**
- **Sterilant**

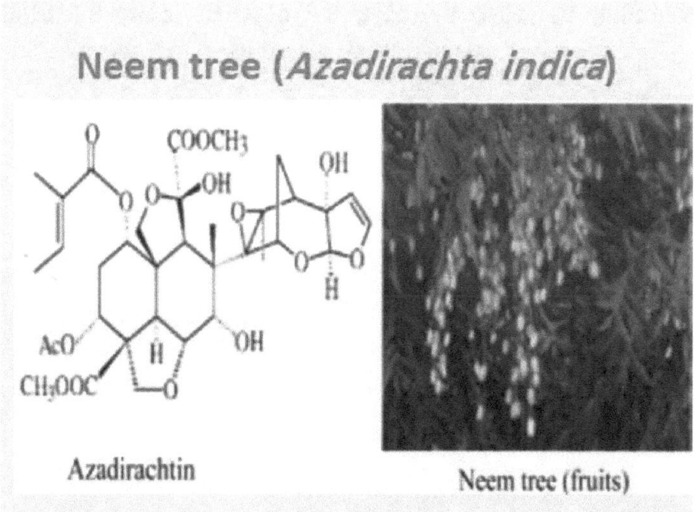

Essential oils:

Advantages of plant extracts

1- The abundance of plants used as plant extracts

2- Most plants have other uses such as medicinal and insect repellent.

3- Its active life is short, which reduces its risks

4- It is used shortly before collecting or harvesting the crop

5- Many of them act as insect repellants

6- Most of this material is more selective for insects, as it has little effect on vital enemies

7-Most of them are not toxic to plants

8- Resistance to these compounds has not been developed as quickly as industrial pesticides.

Defects

1-Most of these products aren't really insecticides because many of them are just insect repellants and have a slow effect.

2- It is rapidly degraded by ultraviolet rays so that its residual action is short.

3- Not all plant pesticides are less toxic to other animals than synthetic ones.

4- It is not necessarily available throughout the season

5- The discrepancy in the proportions of the remaining materials

6-There are no legal records proving its use.

7- Not all recommendations followed by farmers have been scientifically verified.

Carnivorous Plants

- Using carnivorous plants as part of pest control in your greenhouse or glasshouse is a great idea.
- Carnivorous plants often have several attractions such as brilliant colors, sweet secretions and other curious to lure their victims.

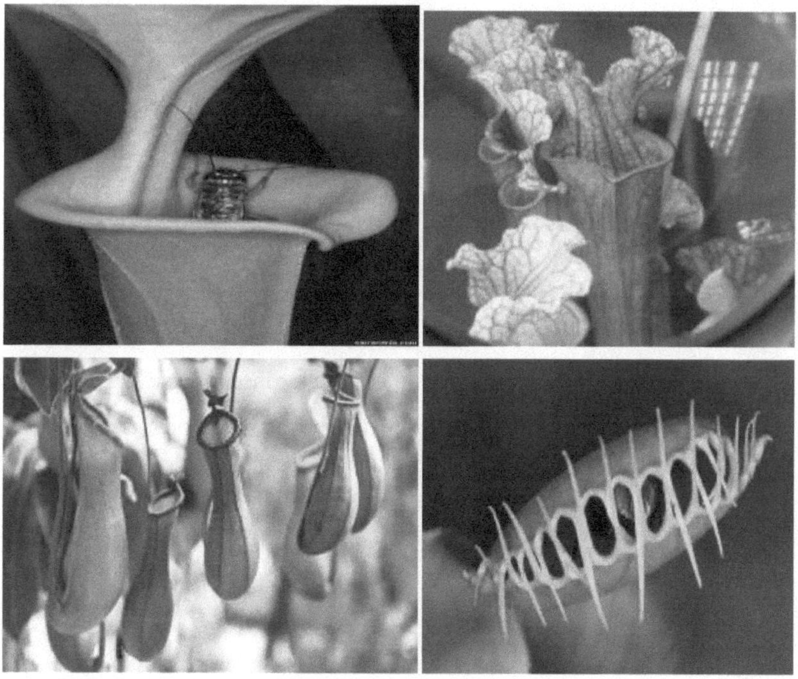

Carnivorous plants are deriving some or most of their nutrients from trapping and consuming insects and arthropods.

Most carnivorous plants have glands that secrete acids and enzymes to dissolve proteins and other compounds. The plants then absorb the nutrients made available from the prey.

Types of insectivorous plants and mechanisms

- Active Traps (Venus Fly trap, *Dionaea muscipula*)
- Semiactive Traps (Sundews, Drosera, Buterworts, Pinguicula)
- Passive Traps (Pitcher plants, Cobralily, Bromeliads)
- **Active Traps (Venus Fly trap, *Dionaea muscipula*)**
- The traps consist of two clamshells like halves, lined with many guard hairs and minute nectar glands.
- A healthy plant may have from 3 to 12 traps.
- Most of the time, for the trap to close, anyone of these trigger hairs must be touched twice, or any two hairs touched one after the other.
- The best closing response seems to happen

Semiactive Traps (Sundews, Drosera, Buterworts, Pinguicula)

These generally have numerous sticky glands covering the upper leaf surface acting like flypaper.

After capture, the leaf margins may role over to cover the prey.

Passive Traps (Pitcher plants, Cobralily, Bromeliads)

Basically tubular and usually topped with a hood.

The creation of nectar lure prey to the pitcher opening.

The slick upper inner lining of the tube causes most insects to fall in to the trap.

There, stiff down ward pointing hairs allow the prey to make a one-way journey deeper in to the ever narrowing tube.

In some species digestive glands secret enzymes which digest the prey. In other species bacterial action breaks down the victim.

The foul smell of the decomposing prey may be helpful in luring other prey.

Bio Pesticides

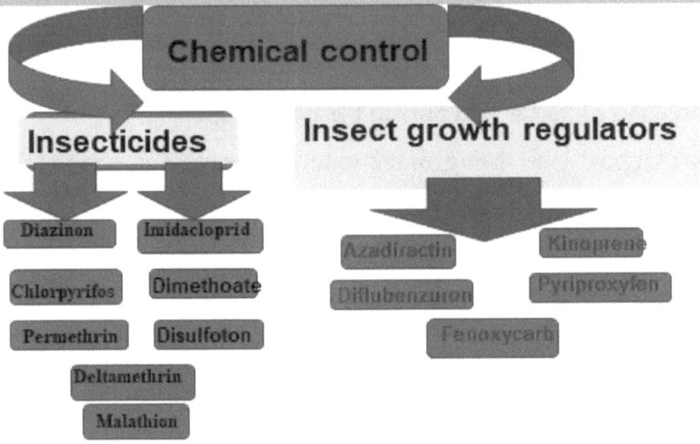

Bio Pesticides

In addition to the above methods, biopesticides are also alternatives to agricultural pesticides. Biocides are a group of pesticides made from natural materials derived from animals, plants, bacteria, fungi and minerals, and are used to control agricultural pests.

There are three types of them:

1- Citronella oil

It is a type of oil extracted from the lemon fruit. It is widely used in agricultural fields to control pests and insects harmful to agricultural crops. This oil is on the US Environmental Protection Agency's (US EPA) list of pesticides with a minimum risk.

2- Garlic oil

Using garlic oil to control agricultural pests According to the Environmental Protection Agency, exposure to some toxic pesticides is a health risk. On the other hand, we find that garlic oil is completely organic. It does not pose any serious health risks, and garlic oil kills unwanted harmful insects.

It has no effect on beneficial insects, helping the ecological balance your garden needs to fend for itself.

3- Acetic acid is an alternative to agricultural pesticides

Acetic acid is a non-toxic organic chemical compound. It is used in many fields, whether industrial, medical and agricultural.

It is an ideal choice for agricultural pest control. Its pungent scent is able to eliminate many pests and insects such as ants, fruit flies, and others.

Some farmers also use it on livestock feed to get rid of bacteria and fungi growth.

How to use biocides

How to use agricultural pesticides Plants, flowers, fruits or herbs are sprayed with garlic spray, acetic acid or citronella oil to repel agricultural pests by spraying the bottoms of the leaves, where insects prefer to lay their eggs.

The plants are sprayed every few days, or once a week, but if you suffer from the problem of a large number of agricultural pests, you should spray daily.

Precautions for the use of bio-pesticides

Although these agricultural pest control materials are non-toxic, they can be irritating to the eyes. It has some effects on the skin, especially if used with additional ingredients such as pepper.

Therefore, be careful while using these materials by wearing some protective equipment while spraying the garden. Also, store these ingredients in the refrigerator for a short period of time if it is not used in time. As these substances can spoil like any other food product.

By reviewing alternatives to agricultural pesticides, we can simply keep the environment clean and pure, as well as preserve human health from any chemical, radiological or other damage.

Safe use of pesticides about new methods of safe control and the search for alternatives to pesticides in order to avoid the harmful effects of their use, such as:

1- The use of nutritional inhibitors, as they work to prevent the insect from feeding, and it will die of starvation.

2- The use of repellents or attractants, which are not toxic and do not lead to the death of the pest, but either attract them to traps or toxic baits, while repellents make the nature of food and the natural life of insects unpleasant, which forces them to flee.

3- Using “pheromones” sexual attractants, which are synthetic materials that carry the scent of females of certain types of pests, in order to attract males. These materials are used either to predict the density of pest populations or to distribute them in the fields for the purpose of dispersing the males of the pest.

4- The use of biological compounds, which are compounds based on microorganisms such as bacteria, fungi or viruses, the most famous of which are bacterial compounds that have entered into practice against many insect pests.

5- Using some alternatives to pesticides (less toxic to plants, animals and the environment), such as mineral oils and plant extracts.

Examples of alternatives to pesticides are:

1- The use of aluminum sulfate (alum saffron) was used to combat the borer and cutworm by making baits and mixing them with half the dose of the pesticide recommended in the control process as an astringent for the jaws and intestines to prevent feeding and eliminate these two pests.

2- Agricultural sulfur use

It has been used to reduce the infestation of sucking insects such as aphids, whitefly, red spider, cotton leaf worm, pink almond worms, thistle and American as a repellent for female butterflies and a killer for newly hatched larvae.

Use of diesel

It has been used to combat the cotton worm and cutworm to kill the larvae and pupae in the soil by adding it to the irrigation water, which leads to the prevention of air oxygen from it, causing its death and elimination.

Use of yeast and molasses

It has been used to combat aphids, whiteflies, scale insects and mealybugs as a disinfectant that competes and eliminates the fungi that grow on honeydew secretions and prevents the appearance of black mold infection

The use of superphosphate and potassium sulfate fertilizer soaked

It was used to reduce the infestation of sucking boring insects such as aphids and whiteflies by changing the pH of the cell juice of the leaves as well as changing the texture of the leaves to a coarse texture that is not suitable for the growth and development of aphids and whiteflies.

The use of return oil (exhaust) and it was used in making grease traps to catch flying insects of aphids, whiteflies and gasids, as well as in resisting fruit tree borers.

The use of neutral soap has been used in spraying against aphids, whiteflies and jassid, to be followed by dusting with sulfur at a rate of 5 kg / acre

From the foregoing, it is noted that the main objective of the process of using alternatives to pesticides is:

Not to interfere with the use of chemical pesticides except in the case of extreme necessity and when the critical limit of infection is reached, at which the damage occurs, with the aim of:

Reducing the total costs used in the control process.

Reducing environmental pollution for humans, animals and plants. This is considered, by the grace and grace of God, a tangible success in preserving the environment, one of the most dangerous and harmful environmental pollutants, namely chemical pesticides.

The advantages of safe alternatives to insecticides are numerous, including:

They are biological compounds and natural materials that are not harmful to humans, plants or the environment.

- Less toxic to pests than chemical pesticides.
- Cheap price for chemical pesticides.
- It begins to be used at lower levels of infestation than chemical pesticides and early detection of infection, so spraying can be repeated to obtain the best results.
- When using biological compounds, the farmer must be confident that the pest will not die immediately, but rather needs an incubation period inside it.
- The grace period after spraying and when picking is almost non-existent if safe alternatives to pesticides are used.
- It is the safe method and suitable for the varying cultural level in the field of pest control
- Errors in using alternatives to pesticides do not cause harm to the farmer, his animals, or his environment.
- Export is one of the basic tasks when applying alternatives to pesticides.
- Repeated use leads to an increase in natural enemies, which reduces the use of chemical pesticides.

- Safe alternatives to pesticides are safe for the product and a guarantee for the source as food is free of chemicals and preserves the environment from pollution.
- Increasing the national and individual product as a result of the success of the control and the enjoyment of human health and wellness.

CONCLUSIONS

There are a number of alternatives to synthetic insecticides available for control of lepidopterous and other pests of crucifers. Some such as the plant based products may be used in a similar way to insecticides but their use still requires careful consideration as some may have detrimental effects on nontarget insects, insect predators other arthropods and fish. It is also sound practice to use when necessary rather than as a preventative measure to maximize efficacy and to delay establishment of resistance.
Some of the options such as biological control with predators and parasitoids involve an acceptance of low levels of the pest larvae to maintain predator populations. The future use of microbial biological control agents will overcome this as they can be applied when and where the pest occurs.

Techniques such as the use of cultivars with some resistance to pests can be adopted with little change to conventional methods of' growing vegetables. Many of the other options, however, are valuable as components of' a pest management program which would involve changes in cultural practices.

Few of the available non-chemical insect control methods have been adopted by farmers (Vereijken 1989). There are a number of reasons for this: some non-chemical control methods are complicated and time consuming being perceived as less reliable, professional and/or effective, and often as more expensive than insecticides.

Some of them require more thought and planning than a regular spray programme. In general chemicals are effective and farmers are accustomed to using them. These reasons are not insurmountable, but probably require a more intensive extension of' research.

There are still many situations where insecticides are presently required: (i) where no alternatives are known or have been tested; and, (ii) in cases of very severe outbreaks. In these instances we should endeavor to use insecticides more efficiently by calibration of spray equipment by correct timing of' applications and by use of' the recommended rate.

Further research is required into developing pest management systems in particular areas using alternatives to synthetic insecticides. Future research into reducing chemical use is likely to involve genetic engineering of plant varieties to enhance resistance to pest attack, development of' biological controls and testing of action thresholds.

REFERENCES

Abivardi C & Benz G. (1984) Tests with the extracts of 21 medicinal plants for antifeedant activity against larvae of Pieris Brassicae L) (Lep Pieridae) Mitteilungen des Sweizerischen Entomologischen Gesellschaft 57, 383 392

Angus T.A. (1968) The use of Bacillus thuringiensis as a microbial insecticide World Review of Pest Control 7, 11-26

Anonymous. 2019. "Agriculture/Biopesticides Market." *Fortune Business Insights* April. https://www.fortunebusinessinsights.com/industry-reports/biopesticides-market-100073.

Australian Bureau of Statistics (ABS). Canberra (1988) summary of Crops Australia 1978-88

Australian Bureau of Statistics (ABS) Canberra (1990). Crops and pastures Victoria 1988-89. Catalogue No. 7321.2.

Baker. G.J. (1984). Damage assessment studies and the derivation of action thresholds for larvae of the cabbage moth. Plutella xylostella L.) on cabbages. Proceedings of the Fourth Australian Applied Entomological Research Conference. Australian Agricultural Council 24-28 September 1984 Adelaide Australia. pp. 175-180.

Baker. R.R. (1970). Bird predation as a selective pressure on the immature stages of the cabbage butterflies, Pieris rapae and P. brassicae. Journal of Zoology. 162 43-59.

Banks. H.J. (1976). Physical control of insects - recent developments. Journal of the Australian Entomological Society, 15. 89-l00.

Bennett P. (1988). Organic Gardening. 4th ed. Child and Associates Publishing Pty Ltd. Frenchs Forest. New South Wales.

Bondarenko' N. (1982). Prospects of biological control of pests in cruciferous vegetable crops in the USSR. Acta Entomologica Fennica. 40. 3-5.

Brunner. E. & Stevens. P.F.E. (1986). The control of diamondback moth with Thuricide. In Diamondback Moth Management pp. 213-217. Asian Vegetable Research & Development Center: Taiwan.

Bryant. I.E. & Yendol. W G (1988). Evaluation of the influence of droplet size and densely of Bacillus thuringiensis against gypsy moth larvae (Lepidoptera: Lymantriidae). Journal of Economic Entomology 81. 130-134.

Buranday R.P. & Rarest R.S. (1975). Effects of cabbage-tomato intercropping on the incidence and oviposition of the diamondback moth Plutella xylostella (L.). Philippine Entomologist 2 369-374.

Bush. G.L. & Neck. R.W. (1976). Ecological genetics of the screwworm fly. Cochliomyia hominivorax (Diptera: Calliphoridae). and its bearing on the quality of mass-reared insects. Environmental Entomology. 5. 821-826.

Cantrell. B.K. (1986). An updated host catalogue for the Australian Tachinidae (Diptera). Journal of the Australian Entomological Society . 25. 255 -265.

Chisholm. M.D.. Steck. W.F.. Underhill. E.W. & Palaniswamy. P. (1983). Field trapping of diamondback moth using an improved tour-component sex attractant blend. Journal of Chemical Ecology . 9 113-118

Chisholm. M.D.. Underhill E.W.. Palaniswamy. Pi & Gerwing V J. (1984). Orientation disruption of male diamondback moths (Lepidoptera: Plutellidae' to traps batted with synthetic chemicals or female moths in small field polls. Journal of Economic Entomology 77 157-160.

Chisholm.M.D.. Underhill.E.W.&Steck.W.F.(1979).Field trapping of the diamondback moth Plutella xylostella using sex attractants. Environmental Entomology. 8. 516-518.

Chow.Y.S..Chiu.S.C &Chien.C.C (1974).Demonstration of a sex pheromone of the diamondback moth (Lepidoptera: Plutellidae). Annals of' the Entomological Society of America. 67 510-512.

Christian. P.D. & Oakeshott. J.G. (1989). The potential of genetically engineered baculoviruses for insect pest control Australian Journal of Biotechnology. 3. 264-266.

Conacher. Jo (1986). Pests. Predators and Pesticides. Some Alternatives to Synthetic Pesticides. Organic Growers Association: Wembley Western Australia.

Cook. I.M.-& Hooper. G.H.S. (1974). Chemoslerilizalion of Plutella maculipennis (Curtis) (Lepidoptera: Plutellidae) by exposure to residues of metepa. Journal of the Australian Entomological Society 13 241-243.

Creighton. C.S.. McFadden. T.L. & Robbins. M.L. (1981). Comparative control of caterpillars on cabbage cultivars treated with Bacillus thuringiensis. Journal of' the Georgia Entomological .Society. 16. 361-367.

Dempster. J.P. (1969). Some effects of weed control ok the numbers of small cabbage white (Pieris rapae (L.)) on brussels sprouts. Journal of Applied Ecology 6 339-345.

Dickson M.H. & Eckenrode C.J. (1975). Variation in Brassica oleracea resistance to cabbage looper and imported cabbageworm in the greenhouse and field. Journal of Economic Entomology 68. 757-760

Eckenrode C J.. Andaloro J.T. & Shelton A.M. (1981). Suppression of lepidopterous larvae in commercial sauerkraut cabbage fields and research plots. Journal of Economic Entomology

Elliott, H.J. & de Little. D:W. (n.d.). Insect Pests of Trees add Timber in Tasmania. Forestry Commission; Hobart. Tasmania.

EPA. 2015. "Active Ingredients Eligible for Minimum Risk Pesticide Products." https://www.epa.gov/sites/production/files/2015-12/documents/minrisk-active-ingredients-tolerances-2015-12-15.pdf.

EPA. 2016. "What Are Biopesticides?" https://www.epa.gov/ingredients-used-pesticide-products/what-are-biopesticides.

EPA. 2019. "Conditions for Minimum Risk Pesticides." https://www.epa.gov/minimum-risk-pesticides/conditions-minimum-risk-pesticides.

Falcon, L.A. (1976). Problems associated with the use of arthropod viruses in pest control Annual Review of Entomology 21, 305-324.

Fisher R & Rosner L (1959) Toxicology of the microbial insecticide Thuricide Agricultural

Fuxa, J.R. (1987). Ecological considerations for the use of entomopathogens in IPM. Annual Review of Entomology. 32. 225-251

Gibson, R.W. & Rice, A.D. (1989). Modifying aphid behaviour. In Aphids. Their Biology Natural Enemies and Control A.K. Minks, & P. Harrewijn, eds.) Vol. C, pp. 209-224. Elsevier;

Glas, P.C.. Smits, P.H . Vlaming, P. & van Lenteren, J.C. (1981). Biological control of lepidopteran pests In cabbage crops by means of inundative releases of Trichogramma species (T evanescens Westwood and T cacoeciae March) a combination of field and laboratory experiments. Medelingen van de Faculteit Landbouwwelenschappen. Rijksuniversitert 46,487-497

Goodwin, S. (1979). Changes in numbers in the parasitoid complex associated with the diamondback moth, Plutella xylostella (L.) (Lepidoptera), in Victoria. Australian Journal of Zoology

Hamilton, JT (1979) Seasonal abundance of Pieris rapae(L.),Plutella xylostella(L.)and their diseases and parasites. Genera/ Applied Entomology 11' 59-66.

Hogan, T. (1990). The dawn of a new era in plant protection. Agricultural Science, 3. 49-50.

Howland. A.F.' Vail, P. & Henneberry, T.J (1966). Results of cage experiments with sterile male releases and a chemosterilant technique for control of cabbage looper populations. Journal of Economic Entomology' 59 194-196.

Hylton, W (ea.) (1974) The Rodale Herb Book. How to use grow and buy nature's miracle plants. Rodale Press Book Division; Emmaus Pa., U S A

Jaques' R.P. (1970). Application of viruses to soil and foliage for control of the cabbage looper and Imported cabbage worm. Journal of Invertebrate Pathology 15 328-340

Jaques, R.P. (1977). Field efficacy of viruses infectious to the cabbage looper and imported cabbageworm on late cabbage. Journal Of Economic Entomology, 70, 111-118.

Jaques, R.P. (1988). Field tests on control of the imported cabbageworm (Lepidoptera: Pieridae) and the cabbage looper The Canadian Entomologist 120, 575 – 580

Jaques' R.P. & Laing J.E (1984). Artogeia rapae (L.), imported cabbageworm (Lepidoptera: Pieridae) Trichoplusia ni (Hubner), cabbage looper (Lepidoptera: Noctuidae) and Plutella xylostella a (L.), diamondback moth (Lepidoptera: Plutellidae). In Biological Control Programmes Against insects and Weeds in Canada 1969-/980 (J.S. Kelleher & M.A. Hulme eds.), pp. 15-18 Commonwealth Agricultural Bureaux, Canada.

Jones, R. E. (1987) Ants' parasitoids and the cabbage butterfly Pieris rapae. Journal of Animal

Jutsum, A.R., Poole, N.J., Powell, K.A. & Bernier, R.L. (1989). Insect controls using micro~a toxins: status and prospects. In Progress and Prospects in Insect Control pp. 129-144 British Crop Protection Council Monograph No. 43

Kenny, G.J. & Chapman, R B (1988). Effects of an intercropp on the insect pests, yield and quality of cabbage. New Zealand Journal of Experimental Agriculture 16, 67-72

Koehler. CS;, Barclay: L.W. & Kretchun, T.M. (1983a). Soaps as insecticides. California agriculture

Koehler I C S.,Barclay, L.W. & Kretchun. T.M. (1983b) Companion plants California Agriculture

Latheef, M.A. & Irwin' R.D. (1979). The effect of companionate planting on lepidopteran pests of cabbage The Canadian Entomologist, 111 863-&64

Latheef,M.A.&Ortiz J.H.(1983a).The influence of companion herbs on egg distribution of the imported cabbageworm, Pieris rapae (Lepidoptera' Pieridae) on collard plants

Latheef' M.A. & Ortiz, J.H. (1983b). Influence of companion plants on oviposition of imported cabbageworm, Pieris rapae (Lepidoptera: Pieridae), and cabbage looper, Trichoplusia ni (Lepidoptera: Noctuidae), on collard plants. The Canadian Entomologist 115,.1529-1531.

Levin, D.B.' Laing, J.E. & Jaques, R.P. (1979). Transmission of granulosis virus by Apanteles glomeratus to its host Pieris rapae. Journal of Invertebrate Pathology 34, 317-318.

Levin, D.B.' Laing, J.E., &Jaques, R.P. (1981). Interactions between Apanteles glomeratus(L.) (Hymenoptera: Braconidae) and granulosis virus in Pieris rapae (L.) (Lepidoptera: Pieridae) Environmental Entomology, 10' 65-68.

Lindquist' D A. (1986). The sterile insect technique and area wide insect control or eradication in the tropics. In Towards a Second Green Revolution. From chemical to new biological technologies in agriculture in the tropics(G.B. Marini-Bettolo ed.), Development in agricultural and managed-forest ecology 19. Proceedings- of the International meeting organised by the Accademia Nazionale delle Scienze detta dei XL' Palazzo Civilta del Lavoro, Roma, Italy in collaboration with the Commission of European Communities and with ENEA. held in Rome, Italy, 8-10 Sep. 1986' pp. 407-414, Elsevier

Little, B. (1989)~ Companion Planting in Australia Reed Books Pty Ltd.: Frenchs Forest, New South Wales.

Lundgren, L. (1975). Natural plant chemicals acting as oviposition deterrents on cabbage butterflies (Pieris brassicae (L.), P. rapae (L.) and P. napi(L.)). Zoologica Scripta 4, 253-258.

Mansour, N.S. (1989) Versatile row and field covers American Vegetable Grower March 1989 72-78

McEwen, F.LM1978). Food production-the challenge for pesticides. Bioscience 28. 773-777

McGaughey. W.H. (1985). Insect resistance to the biological insecticide Bacillus thuringiensis Science 229. 193-195. –

McGaughey. W.H. & Beeman, R.W. (1988). Resistance to Bacillus thuringiensis in colonies of indianmeal moth and almond moth(Lepidoptera: Pyralidae).Journal of Economic Entomology, 81 28-33.

Morallo-Rejesus' B. (1986). Botanical insecticides against the diamondback moth Diamondback Moth Management pp 241-255. Asian Vegetable Research and Development Center; Taiwan.

Morgan, W.C, (1989). Alternatives to herbicides Plant Protection Quaterly/v 4, 33-37.

NPIC. 2020. "Natural and Biological Pesticides." http://npic.orst.edu/ingred/ptype/natbio.html.

Ordish, G.. (1967). Biological methods in crop pest control. Constable and Company Ltd.: London. Great Britain

Parker. F.D. (1970). Seasonal mortality and survival of Pieris rapae (Lepidoptera: Pieridae) in Missouri and the effect of introducing an egg parasite, Trichogramma evanescens~ Annals of the Entomological Society of America 63. 985-994.

Parker, F.D., Lawson F.R. & Pinnell, R.E. (1971). Suppression of Pieris rapae using a new control system: mass releases of both the pest and its parasites. Journal of Economic Entomology), 64, 721-735. ' .

Parker, F.D. & Pinnell. R.E. (1972). Further studies of the biological control of Pieris rapae using supplemental host and parasite releases. Environmental Entomology 1, 150-157.

Parker, F.D. & Pinnell, R.E. (1973). Effect on food consumption of the imported cabbageworm when parasitized by two species of Apanteles. Environmental Entomology 2 216-219.

Philbrick, H. & Gregg, R.B. (1982). Companion Plants and how to Use Them. The Devin-Adair Company: Old Greenwich, Connecticut.

Pimentel, D. (1961). Natural control of caterpillar populations on cole crops. Journal of Economic Entomology 54, 889-892.

Pimentel, D. (1985). Pests and their control: Losses to pests. In CRC Handbook of Natural Pesticides: Methods (N.B. Mandara ea.) Vol. I. Theory, practice and detection, p. 3-19. CRC Press: Boca Raton, Florida.

Pimentel, D., Andow, D., Dyson-Hudson, R., Gallahan, D., Jacobson, S., Irish, M., Kroop' S., Moss. A.' Schreiner.1.'Shepard, M., Thompson, T

& Vinzant' B. (1980). Environmental and social costs of pesticides: a preliminary assessment. Oikos. 34' 126-140.

Pimentel' D. & Edwards' C.A. (1982). Pesticides and ecosystems. Bioscience' 32 595-600.

Radcliffe' E.B. & Chapman' R.K. (1966). Varietal resistance to insect attack in various cruciferous crops. Journal of Economic- Entomology, 59. 120-125.

Rice. M. (1989). Neem seeds. A source of effective and ecologically sound insecticides. Organic Growing. Winter 1989' 21-23.

Richardson R.H., Elliston, J.R. & Avernoff. W.W. North America. Science,1982). Autocidal control of screwworms in North America

Schmutterer H 1990 Properties and potential of natural pesticides from the nee tree, Azadirachta indica, Annual Review of Entomology, 35, 271-297

Sears, M.K., Jaques, R.P. & Laing J.E. (1983). Utilization of action thresholds for microbial and chemical control of lepidopterous pests. (Lepidoptera: Noctuidae` Pieridae) on cabbage Journal of Economic Entomology, 76' 368-374.

Sivapragasam. A.' Ito' Y. & Saito, T. (1988). Population fluctuations of the diamondback moth, Plutella xylostella (L.) on cabbages in Bacillus thuringiensis sprayed and non sprayed plots and factors affecting within-generation survival of immatures. Research on Population Ecology 30, 329-342.

Smith, C.N.' LaBreque' G.C. & Borkovec A.B. (1964). Insect chemosterilants. Annual Review of Entomology' 9' 269-284.

Smith, A.E. & Secoy, D.M. (198i). Plants used for agricultural pest control in western Europe before 1850. Chemistry & Industry, No. 1, 12-17

Smith' C.L. & Smilowitz. Z. (1976). Growth and development of Pieris rapae larvae parasitized by Apanteles glomeratus. Entomologia experimentalis e' Applicata 19. 189-195

Steinhaus~ E.A. (1959). On the improbability of Bacillus thuringiensis Berliner mutating to forms pathogenic for vertebrates. Journal of Economic Entomology, 52 506-508

Steinhaus E.A. (1973). Microbial diseases of insects. In Biological Control of insect, Pests and Weeds (P. DeBach ea.) pp. 515-547. Chapman and Hall Ltd.: London.

Talekar N.S. Lee. S.T. & Huang S.W. (1986). Intercropping and modification of irrigation method for the control of diamondback moth. In Diamondback Moth Management. pp. 145-151. Asian Vegetable Research & Development Center: Taiwan

Tompkins G.J.. Linduska. J.J. Young J..M. & Dougherty E.M. (1986). Effectiveness of microbial and chemical insecticides for controlling cabbage looper (Lepidoptera: Noctuidae) and Imported cabbageworm (Lepidoptera: Pieridae) on collards in Maryland. Journal of Economic Entomology, 79 497-501.

USDA. 2019. "The National List." https://www.ams.usda.gov/rules-regulations/organic/national-list.

van Emden. ELF. (1982). Principles of implementation of IPM. In Proceedings of Australasian Workshop on Development and implementation of IPM (P.J. Cameron C.H. Wearing & W.M. Kain eds.) pp. 9-17. Government Printer: Auckland New Zealand

Vereijken P. (1989). From Integrated Control to Integrated Farming an experimental approach. Agriculture Ecosystems & Environment 26. 37-43

Waters C.. Morgan. W. & Tomkins B. (1988). Crucifer industry statement for Victoria with particular reference to the Gippsland region.

Department of Agriculture& Rural Affairs Marker Report Series No. /1. Agdex 254/853. pp. 33.

Waterhouse D.F. & Norris K.R. (1987). Biological Control. Pacific prospects. Inkata Press Melbourne.

Wilding N. (1986). The pathogens of diamondback moth and their potential for its control--a review. In Diamondback- Moth Management. pp. 219-232. Asian Vegetable Research & Development Center. Taiwan

Wilkinson A.T S. (1966). Apanteles rubecula Marsh and other parasites of Pieris rapae in British Columbia. Journal of Economic Entomology, 59 1012- 1013.

Wilson F. (1960). A review of the biological control of insects and weeds in Australia and Australian New Guinea. Commonwealth Institute of Biological Control Technical Communication I I - 102.

de Wit. H.C.D. (1966). Plants of the World. The HigherPlan~sl. Thames&Hudson Ltd. London.

Woodward P. (1986). An Australian Herbal. Hyland House Publishing Ply. Limited, Melbourne

Xerces Society. 2019. "Comparative Overview: Organic Pesticides." https://xerces.org/sites/default/files/2019-09/19 005_01_Organic-Approved-Pesticides_Overview-FS_web.pdf

Yarrow W.H.T. (1970). Parasites of Plutella xylostella (L.) in south-eastern Queensland Queensland Journal of Agricultural and Animal Sciences. 27' 321 -324

About Author

Prof. Dr. Mohamed Abdel-Raheem Ali Abdel-Raheem

Professor of Entomology (Biological Control)

Pests & Plant Protection Department

Agricultural and Biological Research Institute

National Research Centre, Cairo, Egypt.

33rd ElBohouth St., Dokki, Cairo, Egypt.

Email Address: abdelraheem_nrc@hotmail.com,

abdelraheem_nrc@yahoo.com

Mobile Phone: (+2) 01155527583 - (+2) 01009580797

Prof. Dr. Mohamed Abdel-Raheem Ali Abdel-Raheem, Prof. of Entomology (Biological Control), Pests & Plant Protection Department, Agricultural and Biological Research Institute, National Research Centre, Cairo, Egypt. Published publications (310) papers (83), Books and Book chapters (227), (Scopus) h-index (11), Citations (307), (Google Scholar) h-index (19) , Citations (1094), (Web of Science) Publications (9), h-index (5), Citations (51), Research Gate, h-index (14), Citations (661), Reviewers in International Journals (149), Reviewed articles in

Biological Control (857), Editor in Chief in Journal (7), Assistant Editor (1), Associated Editor in Chief in Journal (5), Editorial board in Journals (36), Advisory board (1), conference Attended (27), workshop Attended (296), Symposium Attended (179), Forum Attended (4), Prize (4), Attended Others (30), Projects As PI and member (19), Training courses For Agric. Engineering (61), Training courses For Students of University (8), attending Trainees courses (19), TV, Radio & Newspaper (30), Supervisor on Ms.C. thesis (1), Supervisor on Ph.D. thesis (1), Committee Ph.D. thesis (2), and member in scientific Foundation (14), Giving International lectures (4).

National Research Centre. 33rd ElBohouth St., Dokki, Cairo, Egypt.

Email Address: abdelraheem_nrc@hotmail.com, ma.abdel-raheem@nrc.sci.eg, abdelraheem_nrc@yahoo.com, profmohamedabdelraheem@gmail.com, abdelraheem196942@gmail.com, Mohamedabdelraheem42@gmail.com, Mabdelraheem69@gmail.com, mohamed196942@yahoo.com, mabdel-raheem@hotmil.com

Mobile Phone: (+2) 01155527583 – (+2) 01009580797

Scopus Author ID : 55220661700

https://www.scopus.com/authid/detail.uri?authorId=55220661700

http://scholar.google.com/citations?hl=en&user=gBtssEgAAAAJ

https://orcid.org/my-orcid?orcid=0000-0001-9240-064X

https://www.webofscience.com/wos/author/searchAbdel-Raheem, Mohamed Abdel-Raheem Ali - Web of Science Core Collection

https://www.researchgate.net/profile/Mohamed_Abdel-Raheem4/publications?sorting=recentlyAdded

http://livedna.org/20.16018

http://www.resarcherid.com/rid/R-6264-2017

https://www.webofscience.com/wos/author/record/R-6264-2017

https://www.facebook.com/share/p/1Piu6xyFmN/

- TV, Radio & Newspaper links

1. https://www.youtube.com/attribution_link?a=NsH2xGv9mY8&u=%2Fwatch%3Fv%3Dsb46rBG8-Mc%26feature%3Dshare
2. https://youtu.be/Ub8djW0tggo
3. https://m.facebook.com/story.php?story_fbid=492907069288338&id=100076011247165
4. https://youtu.be/dOujUAZmTdw
5. https://m.facebook.com/story.php?story_fbid=492907069288338&id=100076011247165
6. https://fb.watch/gPGrDuzZFX/
7. https://www.youtube.com/watch?si=4UfyQaeHVLPiVR5J&v=nXV4NT-no7o&feature=youtu.be
8. https://www.youtube.com/watch?v=sRH44W6TbcQ
9. https://youtu.be/uq9Qj3AEBHk?si=ymaX06huG2AXJESu
10. https://www.facebook.com/share/v/Qg5s6HW3hD8PLBzp/?mibextid=xfxF2i
11. https://el-yom.com/news/archives/1092767
12. https://el-yomnews.com/?p=1156515
13. https://el-yomnews.com/%D8%A7%D9%84%D8%AF%D9%83%D8%AA%D9%88%D8%B1-%D8%B9%D8%A8%D8%AF%D8%A7%D9%84%D8%B1%D8%AD%D9%8A%D9%85-%D9%85%D8%AD%D9%85%D8%AF-%D8%B9%D9%84%D9%8A-

%D9%8A%D8%AA%D8%AD%D8%AF%D8%AB-%D9%84-%D9%84%D9%84/
14. https://el-yomnews.com/?p=1179515
15. https://www.facebook.com/share/v/4MM85wKRPxsYLqTr/
16. http://https//el-yomnews.com/%D8%A7%D9%84%D8%AD%D8%B4%D8%B1%D8%A7%D8%AA-%D9%81%D9%8A-%D8%AE%D8%AF%D9%85%D8%A9-%D8%A7%D9%84%D8%B9%D8%AF%D8%A7%D9%84%D8%A9-%D9%85%D8%AD%D8%A7%D8%B6%D8%B1%D8%A9-%D8%AA%D9%83%D8%B4%D9%81-%D8%AF%D9%88/
17. https://youtube.com/watch?v=F21ziQxdeco&si=LvSUixmnSg4JM1zK
18. https://youtu.be/F21ziQxdeco?si=jyAwbU8dVwkYzRH4
19- https://el-yomnews.com/المكافحة-الحيوية-بالفطريات-حلول-مبتك/
20- https://el-yomnews.com/archives/1296404
21- https://www.facebook.com/share/p/1AC25ue5XG/
22- https://youtu.be/7VItE5kON2g?si=9rMWMUSGk8Csgjfn

TV and Radio

1- Filming in the program (Meeting the Scientists) - Al-Manara Channel, entitled Biological Control of Insect Pests, on 7/2/2013.
2- Filming in the program (Friends of Al-Manara) - Al-Manara Channel, February 2014.

3- A recording of an interview with the program (From the Secrets of the Universe and Life) Radio Sawt Al-Arab, Cairo on 4/26/2015.

4- Filming in the program (Idea for a bobbin) - Nile Cultural Channel under the title "Using entomopathogenic fungi and plant extracts in combating insect pests that infect sugar crops in Egypt" on 11/1/2015.

5- Filming on (This Morning) program - Nile News Channel under the title: Submitting the specifications for the reclamation of one and a half million feddans as a first stage of the reclamation of 4 million feddans on 11/15/2015. https://www.youtube.com/attribution_link?a=NsH2xGv9mY8&u=%2Fwatch%3Fv%3Dsb46rBG8-Mc%26feature%3Dshare

6- Registration in the program (Our Life Between Our Hands) for North Upper Egypt Radio and Radio, presented by Dr. Essam Fathy, 3 episodes, on 11-29-2021, and it will be broadcast on Saturday and Wednesday at 3:30 pm in January 2022

7- Registration in the program (Our Life Between Our Hands) for North Upper Egypt Radio and Radio, presented by Dr. Essam Fathy, 3 episodes, on 11-29-2021, and it will be broadcast on Saturday and Wednesday at 3:30 pm in January 2022

8- Registration in the program (Our Life Between Our Hands) for North Upper Egypt Radio and Radio, presented by Dr. Essam Fathy, 3 episodes, on 11-29-2021, and it will be broadcast on Saturday and Wednesday at 3:30 pm in January 2022

9- Registration in "Our Green Land" program for North Upper Egypt Radio and Radio, presented by Dr. Essam Fathy, 3 episodes, on 11-2021-2021, to be broadcast in January 2022

10- Registration in "Our Green Land" program for North Upper Egypt Radio and Radio, presented by Dr. Essam Fathy, 3 episodes, on 11-2021-2021, to be broadcast in January 2022

11- Registration in "Our Green Land" program for North Upper Egypt Radio and Radio, presented by Dr. Essam Fathy, 3 episodes, on 11-2021-2021, to be broadcast in January 2022

12- Filming in the program (Egypt is Beautiful) - the second Egyptian Nile Channel, under the title: Climate Changes on November 6, 2022. https://youtu.be/dOujUAZmTdw

13- Registration in "Our Green Land" program for North Upper Egypt Radio and Radio, presented by Dr. Essam Fathy, 3 episodes, on 11-2021-2021, to be broadcast in January 2022

14- Filming in the program (Egypt and Climate Summit) - Delta Satellite Channel under the title: Climate Cop 27 on 13/11/2022 https://youtu.be/Ub8djW0tggo

15- Filming in (Egypt) - Sohag, Coral Times Hotel, under the title: Climate Change, Cop 27, dated 11/15/2022, the journalist Shorouk Abdullah
https://m.facebook.com/story.php?story_fbid=492907069288338&id=100076011247165

16- Filming in (Feed and Benefit) - Coral Hotel in Sohag under the title: Climate Cop 27 on 11/15/2022, journalist Ahmed Tarboush https://fb.watch/gPGrDuzZFX/

17- An intervention in the plant clinic program on insect-pathogenic fungi in Egypt's agricultural farms on Thursday 29/2/2024.

https://www.youtube.com/watch?si=4UfyQaeHVLPiVR5J&v=nXV4NT-no7o&feature=youtu.be

18- Filming in the plant clinic program on the Misr Agricultural Channel about insect-pathogenic fungi on Sunday 3-3-2024

https://www.youtube.com/watch?v=sRH44W6TbcQ

19- Filming in the plant clinic program at the Misr Agricultural Channel, early prediction of the red palm weevil and modern methods of combating it, on Thursday 25/4/2024.

https://youtu.be/uq9Qj3AEBHk?si=ymaX06huG2AXJESu

20- Filming in the Green Planet program on the Egyptian Channel 2 (Biological Control) on Sunday 7/6/2024

https://www.facebook.com/watch/?v=981454710052376

21- A press interview in Al-Youm newspaper with journalist Mr. Imad Naseer on 7-30-2024. https://el-yom.com/news/archives/1092767

https://el-yomnews.com/

22- https://el-yomnews.com/?p=1156515

23- https://el-yomnews.com/%D8%A7%D9%84%D8%AF%D9%83%D8%AA%D9%88%D8%B1-%D8%B9%D8%A8%D8%AF%D8%A7%D9%84%D8%B1%D8%AD%D9%8A%D9%85-%D9%85%D8%AD%D9%85%D8%AF-%D8%B9%D9%84%D9%8A-%D9%8A%D8%AA%D8%AD%D8%AF%D8%AB-%D9%84-%D9%84%D9%84/

24- https://el-yomnews.com/?p=1179515

25- https://www.facebook.com/share/v/4MM85wKRPxsYLqTr/

26- Insects in the Service of Justice: A Lecture Explaining the Role of Forensic Entomology in Investigating Crimes

https://el-yomnews.com/archives/1243894

27- Human health is at risk. An investigation reveals the secrets of pesticide-contaminated agriculture.
https://el-yomnews.com/archives/1296404
برنامج الكوكب الاخضر (الحشرات النافعه ودورها بالبيئة) فى 23-5-2025 القناة الثانية المصرية

28- https://www.facebook.com/share/p/1AC25ue5XG/
The Green Planet Program (Beneficial Insects and Their Role in the Environment) on 5/23/2025 on the Egyptian Channel 2

29- https://youtu.be/7VItE5kON2g?si=9rMWMUSGk8Csgjfn

More Books!

yes

I want morebooks!

Buy your books fast and straightforward online - at one of world's fastest growing online book stores! Environmentally sound due to Print-on-Demand technologies.

Buy your books online at

www.morebooks.shop

Kaufen Sie Ihre Bücher schnell und unkompliziert online – auf einer der am schnellsten wachsenden Buchhandelsplattformen weltweit! Dank Print-On-Demand umwelt- und ressourcenschonend produzi ert.

Bücher schneller online kaufen

www.morebooks.shop

info@omniscriptum.com
www.omniscriptum.com

OMNIScriptum

FSC
www.fsc.org
MIX
Papier | Fördert
gute Waldnutzung
FSC® C083411

Kolibri 360 GmbH
Ferdinand-Jühlke-Straße 7
99095 Erfurt, Deutschland
produktsicherheit@kolibri360.de

Druck:
CPI Druckdienstleistungen GmbH
im Auftrag der
Kolibri 360 GmbH
Ferdinand-Jühlke-Str. 7
99095 Erfurt